Building for Energy Independence

# Sun/Earth Buffering and Superinsulation

Building for Energy Independence

# Sun/Earth Buffering and Superinsulation

**Don Booth with Jonathan Booth and Peg Boyles**

**Design by Ned Williams**
Illustrations and charts by Donald B Johnson
Architectural plans, sections, details by Donald A King

Library of Congress Catalog Card Number 83-72283
ISBN 0-9604422-3-5 Paperback
ISBN 0-9604422-4-3 Hard Cover

For their advice, inspiration, assistance at some
time in the long evolvement of this book,
appreciation to: Bruce Anderson, Tamil Bauch,
Galen Beale, Jim Berk, Helen Brigham, Bill
Brodhead, Bing Chen, Douglas Clayton, Rick
Cowlishaw, Sara Cox, Jane Doherty, Howard M
Faulkner, John Flippen, Drew Gillett, Steve J
Heims, Douglas Holmes, Hank Huber, Bruce
Kaufman, Eugene Leger, Bill Mead, Harold Orr,
Paul Peters, Norman B Saunders, William A
Shurcliff, Pamela Smiley, Skeet Stokes, Sonia
Wallman, and Chris Williams.

And to these other Community Builders: Steve
Booth, David Buzzell, Seny Bynum, Tory Dodge,
Peter Lawless, Bruce Pearl, Dana Pearl, Rodney
Pearl, and especially Lois Booth.

**Printed by The Book Press, Brattleboro, Vermont, USA**
**(Cover by New England Book Components, Hingham, MA)**

Published by

**Community Builders**
Canterbury, New Hampshire 03224

2 4 6 8 10 9 7 5 3

# CONTENTS

# 1

## INTRODUCTION

The last half-dozen years have seen the rapid development of new approaches to building design and construction which use sun, earth, and super-insulation to create buildings that use 60 to 90 percent less purchased energy for heating than most homes built previously.

It is crucially important that those who plan and build houses — owners, designers and builders — know of the tested practicality, comfort, and attractiveness of these new systems. It is an economic and ecological tragedy when cold-climate home-builders of today build with the low standards of energy conservation that characterized most homes of the past.

In the field of energy-efficient house design there is now a large body of real experience which anyone who is planning a new home can wisely draw from. Many designs have been thoroughly tried and proven, and others have been tried and found wanting.

Of the various approaches which have been taken to energy-conserving housing, there are two that we feel stand out as offering maximum satisfaction and practicality. These are the SUN/EARTH BUFFERED HOUSE, an improved outgrowth from the "envelope" house, and the SUN-TEMPERED SUPERINSULATED HOUSE.

Both of these designs, with very little trouble or maintenance, will provide a degree of energy independence in heating needs that will enable their owners to face with confidence the uncertainties of energy supplies and costs in the decades ahead. In New Hampshire and other even colder climates, we have seen these design and construction methods produce homes that are practical, comfortable, and aesthetically pleasing, typically needing no more than $50 to $200 of purchased fuel per year. The intent of this book is to explain clearly, usefully, and succinctly how this can be accomplished.

## The Sun-Tempered Superinsulated House

Of all the approaches to energy-efficient housing, superinsulation is perhaps the most simple and direct: keep the heat loss from the house so low, by using thick insulation and controlling air leaks, that very little heat is required to maintain comfortable indoor temperatures even in zero-degree weather.

With such a low rate of heat loss, the heat which is normally generated within an occupied house — from cooking, appliances, television, people's body heat, etc. — can meet a large portion of a home's heating needs. A variety of new and economical systems have now been developed for building houses with thick, high R-value walls and excellent air and vapor barriers.

This approach is most effective when combined with sun-tempering: placing more windows on the south than on other sides, for moderate warmth from the sun.

Superinsulated homes can be built in any style and in any location. Even if no sunshine is available they still keep fuel needs very small. They are relatively simple to build and very simple to manage.

## The Sun/Earth Buffered House

The creation of dramatically new and useful systems is often the result of a simple reordering of conventional patterns of thought, a rearrangement of familiar components of existing systems. In this manner, an important new building system came into being when several already familiar elements of energy-conscious housing design were combined in a new way. The result is an exciting new kind of human shelter, nearly self-reliant thermally and able to provide its inhabitants with a range of benefits not collectively possible in other shelter designs.

We are calling this concept the SUN/EARTH BUFFERED HOUSE, or more briefly the BUFFERED HOUSE. This name conveys what the house is and suggests how it differs from previous shelter designs. It also suggests what the house can do, namely, buffer its occupants in important physical, economic, and psychological ways.

A buffer moderates between two greatly differing substances or conflicting forces, lessening the impact of their direct encounter. Any house, from cave to castle, is a kind of buffer, protecting its occupants physically from the full force of the out-side elements, and shielding them psychologically by providing a familiar, personal space away from the larger world.

In a buffered design, a part of the house itself becomes a protected intermediate environment between the inner house and the outdoors. This concept received much attention in the "envelope" houses which began appearing in the late 1970s. A buffered house uses key elements of the envelope houses: a south-facing sunspace connected by an air passage to the earth mass underneath the house, heat transfer by moving air, and insulation. We are using the term to include envelope houses, but also to include recent and future designs which use the same key principles but which would not be described as envelope houses (for example, many buffered houses do not have a double north wall).

The sunspace in a buffered house protects and warms the inner house, but more than that it is itself a useful, much-enjoyed space. It may be used as a working greenhouse, designed for maximum plant growth and food production, or it may be intended simply as a pleasant solarium, decorated with plants and used for sitting, eating, children's play, or for an entryway. Often the design will have some aspects of each approach. In any case, the greenhouse or solarium is a focal point of the house and a daily source of enjoyment.

The instability of our fossil fuel supply has been well-known since the energy crisis of 1973. There is also a growing awareness among many people of the fragility of much of our nation's food supply system. Many of our primary food-producing areas are originally deserts and rely on constant irrigation for their productivity, in regions which are increasingly short on water. Our food processing and distribution network is also highly energy-dependent. Year-round home food production and bulk storage of seasonal crops are becoming increasingly popular solutions for those who are aware of the vulnerability of their food supply.

A buffered design can address the dual needs of energy-independence and home food production in an integrated system which goes far beyond just a shelter from the world and weather. The greenhouse which is a part of the design is protected from freezing without requiring auxiliary heat, and creates a delightfully warm space on sunny days, even in the cold of midwinter.

Food and shelter are not the only requirements for human survival. We live in times of enormous uncertainty, of increasing technological complexity and accelerated economic and social change, all of which take their toll on our psychological well-being. To live reasonably well-balanced and productive lives amidst such turbulence, people must create buffers of certainty, beauty, simplicity, durability, and self-reliance. A sun/earth buffered house provides many of these kinds of buffers and itself becomes a personal life-support system.

## About This Book

This book consists of four main parts: the theory section (Chapters 2 and 3), the "how to" of design and construction (Chapters 4-7), the experience with existing buffered homes (Chapters 8 and 9), and the Reference Section. Reference material includes: abbreviations, glossary, related reading, sources of products, greenhouse food production suggestions, table of heat from various fuels, and design calculation information.

The book does not have to be read from front to back, but we do recommend reading the next chapter, "Background and Development," before jumping ahead to the rest of the book. It contains a concise, easy-to-follow orientation to the principles and practices of energy-efficient housing; it provides the tools for understanding and utilizing the remainder of the book; and it acquaints readers with our perspectives and assumptions.

# 2

## BACKGROUND AND DEVELOPMENT

### HOME HEATING
#### Comfort

The fundamental reason people live in houses is to be more comfortable and safe than they would be outdoors; houses buffer their inhabitants from weather extremes. Designers of warm-climate housing can concentrate primarily on comfort alone, but in cold climates a warm shelter is a survival necessity.

Thermal comfort is not governed by temperature alone. Humidity, air movement, the temperature of surrounding surfaces, people's activity levels, and differing perceptions of identical conditions all influence human awareness of thermal comfort.

However, most people feel comfortable sitting lightly clothed in still air at temperatures between 68F and 80F with a relative humidity near 50 percent. Within this range the human body most easily maintains its own internal temperature; this is the range where people generally want their daytime living spaces to be.

#### Heat

Heat is simply the energy of molecular activity, driven always from warmer to cooler, towards uniformity and molecular rest. The greater the difference in temperature between two surfaces or areas, the more rapid the flow of heat from the warmer part to the cooler part. Conversely, with a smaller temperature difference the flow of heat will be slower.

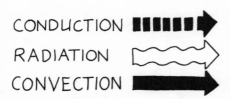

## Types of Heat Transfer

There are three basic ways in which heat is transferred: CONDUCTION, RADIATION, and CONVECTION. In all cases the natural tendency of heat is to move from warmer objects or spaces to cooler objects or spaces.

**Conduction** is the transfer of heat through a material, for example through the walls of a building. The molecular activity that heat represents is transferred directly in and through the substance, always tending toward equal activity and temperature.

**Radiation** is energy transmitted directly through space. Solar radiation passes through the air, to heat and light objects that it strikes. In addition, any object radiates energy which heats any cooler objects around it. As an example, a person in a room with cold walls will radiate heat to them, and can feel cool even if the air in the room is warm.

**Convection** is the transfer of heat by the movement of fluids such as air or water. For example, the heated air around a stove will tend to rise to the ceiling. That buoyancy is one form of "natural" convection; "forced" convection refers to the use of fans or pumps to move a fluid and the heat contained in it.

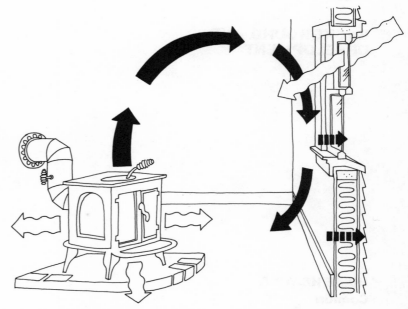

One of the greatest challenges in house design is providing a comfortable interior in a climate where outside weather conditions range far from human comfort levels. Throughout this chapter and most of the book we focus attention primarily on housing design for cold climates; however, many of the same design features that keep a house warm during cold winters will also keep a house cool in hot weather.

There are two basic strategies for maintaining cold-weather comfort levels inside a building. The first is simply to add enough heat to the interior to replace the heat being lost to the outside. The second approach is conservation: minimize heat losses by using building materials that impede the flow of heat from the structure.

Until recent decades, not much attention was given to controlling heat losses from American houses. The earliest interior heat sources were wood-burning fireplaces which provided radiant heat to people huddled nearby, but did little to keep the rooms warm. What little convected heat had not been sucked out by greedy chimneys was lost rapidly through cracks and uninsulated walls. Without insulation, rooms were drafty and unevenly heated, long after the invention of more efficient box stoves and coal-burning appliances.

During the era ot cheap and abundant petroleum, the typical American house had an inexpensive, heat-leaking shell and one or more heating appliances sized more than large enough to replace the heat flowing rapidly and continuously out through the structure during cold weather. To maintain relatively uniform temperatures throughout the house without putting a separate heating appliance in every room, large central furnaces were installed in the cellar and equipped with fans or pumps to move the heat into the various rooms.

Insulating and caulking materials and infiltration barriers that resist heat loss did not come into widespread use until around the middle of this century, and then not primarily as conservation measures, but as means of preventing drafts and providing more even temperatures throughout the house.

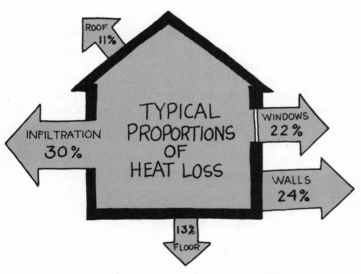

Typical proportions of heat loss from an insulated house with a moderate amount of windows. The arrow for wall heat loss represents all the walls, the window arrow represents all the windows, etc. The percentage of heat loss due to infiltration can vary from under 20 percent to more than 50 percent. In a one-story house the roof heat loss percentage would usually be somewhat greater.

The heat losses of a house are primarily by CONDUCTION and INFILTRATION.

**Conduction Heat Loss**

Conduction losses result from the transfer of heat directly through the materials of the building shell. The amount of heat lost or transferred through a material by conduction is determined by three factors:

1. the RESISTANCE, or R-value, of the material;
2. the TEMPERATURE DIFFERENCE, or Delta T, between the inner and outer surfaces of the material; and
3. the AREA over which the loss is taking place.

To calculate the amount of heat (per hour) lost through a material by conduction, divide the area by its R-value and then multiply by the temperature difference.

Heat lost per hour =
(area ÷ R-value) × (Inside T — Outside T)

Note that if the outside temperature is greater than the inside temperature the result will be negative, signifying the amount of heat GAINED from the outside.

Assuming conventional units with temperatures in degrees Fahrenheit, the result of the above equation is in British thermal units, or Btus, per hour.

The key point is that conduction loss is DIRECTLY proportional to the temperature difference (the Delta T), and INVERSELY proportional to the R-value.

This means that the conduction loss through a square foot of an R-2 window (double-glazed) is ten times the loss through a square foot of an R-20 wall (6″ wall insulated with fiberglass). In addition, the loss from a 70 degree interior temperature to a 10 degree outdoor temperature (Delta T = 60) will be twice as great as the loss from a 70 degree interior to a 40 degree outdoor (Delta T = 30), all other conditions being equal.

Resistance (''R'' or ''R-value'') is a measure of a material's ability to impede the flow of heat through it. Resistance is actually the inverse of conductance (''U'' or ''U-value''), the measure of a material's ability to conduct heat through it. In other words:

$$R = 1 / U$$
$$\text{and} \quad U = 1 / R$$

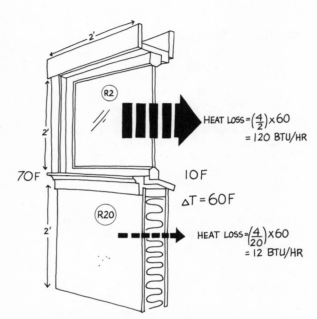

$$\text{HEAT LOSS} = \left(\frac{4}{2}\right) \times 60$$
$$= 120 \text{ BTU/HR}$$

70F                     10F

$$\Delta T = 60F$$

$$\text{HEAT LOSS} = \left(\frac{4}{20}\right) \times 60$$
$$= 12 \text{ BTU/HR}$$

Heat loss through four square feet each of a window and a wall. Heat is going from the 70F interior (left side of drawing) to the 10F outdoors, a temperature difference or Delta T of 60F. Heat loss through the R-2 window is ten times the heat loss through the R-20 wall.

## Infiltration Heat Loss

Infiltration is the process of (cold) outside air entering a house and replacing (warm) inside air leaving it. Substantial heat loss can result from the infiltration in a "loose" house, i.e. a house with many unstopped cracks. In fact, the efforts put into insulating a house very well can be largely wasted if no attempt is made to reduce air infiltration. That is why a crucial element of superinsulated construction is a very good infiltration barrier.

Infiltration is commonly described in units of "air changes per hour" (AC/hr), which is the number of times in one hour that a volume of air equivalent to the total house volume enters from outdoors. The amount of heat lost by infiltration is determined by the volume of air exchanged — the air change rate — and by the temperature difference between the inside and outside of a house.

Note that both conduction and infiltration losses are directly proportional to the temperature difference between inside and outside; in both cases it is the temperature difference that causes the flow of heat.

Much later, the oil embargo of 1973-1974 brought wide recognition that energy from fossil fuels would never again be cheap, and sharply increased public awareness of the economic value of minimizing heat loss from buildings. Insulating and tightening a new or older house is seen today as the single most cost-effective way homeowners can reduce present and future energy costs.

## Sunshine Replacing Oil

It was natural for house designers to look for ways of using sunshine more effectively for space heating. The sun is the earth's primary heat source. Its heat cannot be metered, shut off, or interrupted as leverage during international political disputes. As a raw source of heat sunshine is essentially free, and in providing heat to buildings it does not add any pollutants to the air, water, or soil. By the late 1970's, energy-conscious designers had developed numerous innovative ways to use the sun's energy in designs that use far less purchased energy for heating than conventional houses.

In reviewing some of these basic developments it is worth noting that solar house design has not evolved according to some master plan, but rather mostly by individual trial-and-error efforts. Technical conferences and personal communications among designers and builders, performance data from monitored houses, and media attention have all contributed to the flow of information among interested people.

There is still much to be learned about solar design, but perhaps the largest problem is that cutting-edge information trickles out slowly to the people who need it the most. Many homes are still being built with little awareness of energy-efficient techniques, and design errors of some of the first solar homes continue to be repeated.

The global energy crisis also made it imperative for every petroleum-consuming sector of society to look for alternative sources of energy.

## MAJOR TYPES OF ENERGY-SAVING HOUSES
### Direct Gain Solar Homes

People have always invited sunshine into their homes to provide daylight and a sense of well-being. Most people have experienced how warm a sunlit room can become, even on a cold winter day. This is largely because of the phenomenon known as the "greenhouse effect" created by glass and other glazing materials. In many conventional houses designed to admit much light, people must close shades or open doors or windows on sunny days even in cold weather to prevent the rooms from overheating.

The simplest solar heating systems, called DIRECT GAIN systems, make conscious use of this principle. Direct gain houses are oriented so that large window areas face south to collect the most sunlight. To prevent summer overheating, windows usually have been designed with a shading overhang to exclude the high arc of the summer sun while admitting the low arc of the sun in winter.

Effective direct gain houses also employ good insulation at walls, roof, foundation, and at windows at night, to ensure that the collected solar heat does not flow immediately back out of the building. Windows are poor insulators, and large glazed areas that collect heat while the sun is shining are also vulnerable to large heat losses at night and on cloudy days.

### Greenhouse Effect

Anyone who enters a closed car that has been sitting in the sun experiences at first hand the "greenhouse effect." Solar radiation is primarily in the range of visible light, and readily passes through transparent glazing and is absorbed by objects inside and converted to heat. These sun-warmed objects slowly re-radiate their heat to the rest of the room and to the glass and outdoors. Much of the inside build-up of heat from the sunshine is due to the protection the enclosure provides against heat loss by convective air movement.

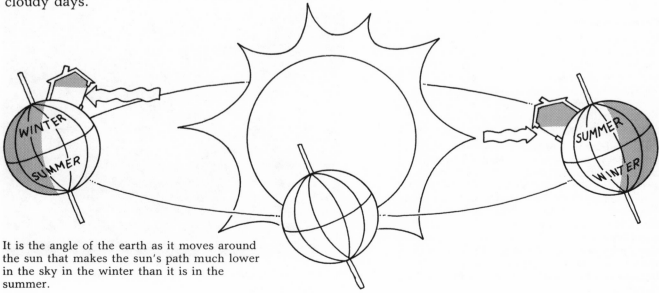

It is the angle of the earth as it moves around the sun that makes the sun's path much lower in the sky in the winter than it is in the summer.

### Sun Angles

One of the conditions that make solar energy practical in simple systems for heating houses is the difference between the angle of the sun in winter (when we need its added heat) and the angle in summer (when we do not). Much more solar radiation is transmitted through simple vertical south glazing in winter than in summer. Deciduous trees and proper overhangs can accentuate this seasonal difference.

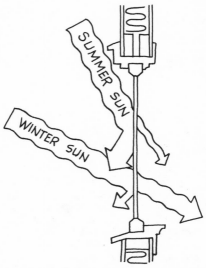

Summer and winter sun coming through a south window

### Thermal Storage

The heat storage ability of an object is called thermal capacitance or capacity, and is a function of the material's mass and its specific heat. Thermal capacitance determines how slowly an object changes temperature when a given amount of heat is supplied or removed. (Various materials' capacitances are listed in the Glossary in Reference Section B.)

The actual amount of heat contained (stored) in a material at any time is determined by both capacitance and temperature. At the same temperature, a material with large capacitance contains more heat than a material of the same volume but of smaller capacitance. Low-grade heat refers to heat at a relatively low temperature, even though the actual quantity of heat stored may be quite large. High-grade heat refers to heat at higher temperatures, usually meaning a temperature well above the assumed human comfort range.

The single greatest challenge in using the sun for space heating is in providing heat for nights and sunless periods. This is especially a challenge in northern areas that have short days, long cloudy spells, and high heating requirements.

All effective direct gain systems require some sort of THERMAL STORAGE capable of absorbing large quantities of heat while the sun is shining and releasing the stored heat into the living space when the sun is not shining. Containers of water placed where they will receive direct sunshine are a popular means of storing solar heat: water is cheap, abundant, and has the highest heat storage capacity of any common substance. Rock, cement, brick, and other masonry materials are also effective heat storage materials. They do not store as much heat per unit volume as water, but they have the advantage that they can serve a dual function as structural materials which will not occupy additional interior space. Concrete floors and masonry walls or chimneys are frequently used to incorporate significant added thermal storage in a structure.

Thermal storage systems not only store heat for sunless periods; they also prevent the overheating that would otherwise occur on sunny days with large south glazing areas. In this way thermal storage acts as a buffer against wide interior temperature swings. Well-designed direct gain systems are based on a carefully calculated balance

The floor and these walls are concrete in this direct gain house.

between glazing area and thermal storage capacity. Poorly designed direct gain systems can overheat on sunny days and cool rapidly during sunless winter periods.

One important variation of the direct gain concept uses internal convective loops to distribute sun-heated air throughout the house. This results in more even temperatures and more efficient thermal storage since heat is stored throughout more of the structure.

Direct gain systems are simple in concept. They can be inexpensive and effective, and they contain no moving parts that need maintenance and eventual replacement.

They do have some disadvantages. The glare through large glass areas can be intense, creating an uncomfortable space behind the south glazing. The sunlight can fade rugs, upholstery, and drapes. Using a light-diffusing glazing material will reduce glare and fading, but will also impede vision to the outside. In heavily populated areas, large south windows may afford passers-by too intimate a view of the home's interior.

Some people find the thermal storage systems of direct gain houses obtrusive and unpleasant, and are unwilling to live on concrete floors, surrounded by concrete walls or walls of water storage containers. The lack of visual and acoustical privacy created by the floor plan in some direct gain houses can be unacceptable. Because of their massive heat storage systems, direct gain houses may be slow to heat up when they are cold and slow to cool down if they overheat. In cold climates they can still require considerable amounts of supplemental heat.

For the best thermal performance in direct gain houses, night insulation for windows is usually essential, but movable window insulation systems can be expensive, and require attention and effort that many homeowners may not want to give.

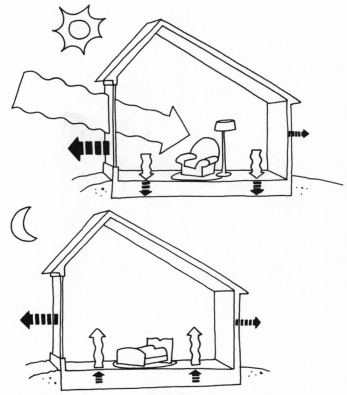

**Direct Gain House, Day and Night**

In the daytime, sun entering the house through large south windows is absorbed in the thermal storage mass (in this case, the concrete floor). In the nighttime the thermal mass helps keep the living space warm.

**Direct Gain House with Internal Convective Loop**

Convective air movement has been encouraged in this direct gain house by the placement of air openings.

These thermosiphon air-heating panels circulate air through hollow-core concrete floors.

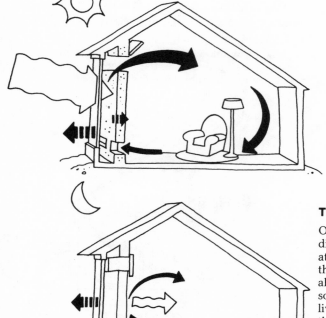

## Indirect Gain Systems

Indirect gain solar heating systems overcome some of the limitations of the direct gain designs. Although there are a number of quite different indirect gain strategies, they all separate the heat collection system and sometimes the heat storage system from the living space itself.

The use of an indirect gain system may be particularly appropriate when a large direct outlook to the south would be undesirable for reasons of privacy or because of an unattractive view or noise.

One simple indirect gain system is the thermosiphon air-heating panel. This is a south-facing panel with glazing on the outside, a dark, heat-absorbing surface on the inside, and an airspace between. Room air enters through an opening at the bottom of the panel, grows warmer and travels upward by natural convection, and re-enters the room through an opening at the top. Thus, whenever solar radiation is warming the panel, a continuous flow of warmed air is entering the room.

Another simple and popular indirect gain system is the "Trombe" wall. A massive heat-storing wall, usually of concrete or other masonry, is built into a structure's south face, set behind an extensive glazed area where it collects the sun's heat. On sunny days, some of the heated air behind the glass may move by convection through openings in the wall directly into the living space. The massive wall itself absorbs heat during the daylight hours and radiates some of that heat slowly out the other side into the living space at night.

## Trombe Wall, Day and Night

On sunny days the massive concrete wall directly behind south glazing absorbs heat, and at night the heat stored in the wall helps keep the interior warm. The Trombe wall shown here also includes thermosiphon vents, which allow some of the heated air to circulate into the living space on sunny days. It is important that the thermosiphon vents not allow reverse circulation at night, with cool air falling down behind the glazing and entering the living space through the bottom vent.

The popular attached solar greenhouse is an indirect gain heating option which can incorporate a winter food-production capacity. The greenhouse features an expanse of south glazing for heat and light collection, and usually considerable thermal storage capacity to reduce overheating on sunny days and prevent freezing on cold nights. Excess heat not needed by greenhouse plants can be used for heating the living space by opening doors or windows adjoining the greenhouse.

A well-managed solar greenhouse can provide a family with a supply of fresh, nutritious greens all winter, as well as a modest amount of solar heat to help cut heating costs, but its popularity is perhaps due mostly to the fact that the greenhouse becomes a plant-filled refuge in midwinter. Any veteran of a few northern winters can appreciate the rich psychological benefits of a sun-drenched greenhouse. The attached solar greenhouse is an option which is particularly well-suited to retrofitting on an existing house.

Indirect gain solar heating systems usually do not provide a large proportion of a home's heating needs in a cold climate. A greenhouse which has to store enough heat when the sun shines to keep plants from freezing at night can spare less heat for the house. Any heating system in which the massive materials for heat storage are directly behind large glazed areas has an inherent inefficiency, since the stored heat tends to go right back out through the glass.

## Active Systems

The solar heating strategies discussed so far have been PASSIVE systems that rely solely on natural processes of convection, conduction and radiation for collecting and distributing the sun's heat within the building. But for several years after 1973, by far the largest number of domestic solar space heating units built were ACTIVE systems.

Active solar heating systems are indirect gain devices in which collected solar heat is transferred to a working fluid — usually air or water — and then mechanically pumped or blown into a thermal storage medium. During sunless periods, the stored heat is pumped out into the living space. Many homes incorporate both active and passive solar

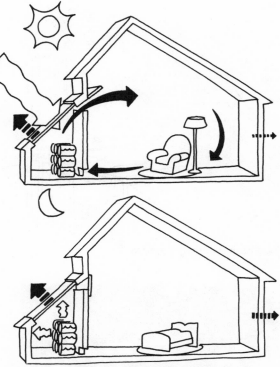

**Indirect Gain Attached Greenhouse, Day and Night**

During the day, heat that is not absorbed in the water barrels is used to warm the interior of the house. At night, the greenhouse is closed off from the livingspace and the water barrels keep the greenhouse from getting too cold for plants.

Waterbeds in this attic absorb solar energy through a glazed south roof. A fan blows warm air from the heavily-insulated attic to the rooms below.

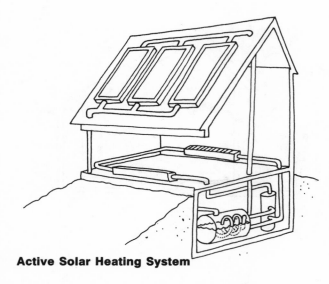

**Active Solar Heating System**

House by Don Metz

heating components, receiving solar heat passively by direct gain or through a greenhouse on sunny days and employing an active rooftop collector to pump heat into storage for use at night and on cloudy days.

Active systems typically permit a more flexible house layout and orientation, since both heat collection and storage can be remote from the living space. Active systems may be most appropriate for retrofitting to existing dwellings. The use of forced convection permits rapid collection and allows the stored heat to be sent only where and when it is called for.

Active systems are generally more expensive to buy than passive systems. They require expert installation and maintenance, and contain moving parts that may need replacement. They depend on electricity and are vulnerable to power outages that can occur at the peak of the heating season. Their added-on appearance may be considered unattractive. In recent years most attention has turned away from active systems to simpler passive systems, although small fans are often used to circulate air in otherwise passive systems.

## The Underground House

Homebuilders in both hot and cold climates have long been attracted to the idea of putting a house into the ground, nearer to the steady temperatures of the deep earth. Deep earth temperatures are closer to human comfort levels than are the extremes of outside weather, so an underground house can require far less energy than a conventional house to keep its occupants comfortably warm or cool — especially if an exposed side faces south.

The design of an underground house requires careful engineering and a high material cost to support the heavy loads of an earth roof — plus rain and snow — and to achieve dependable waterproofing. To provide daylight and egress, the arrangement of rooms must be either linear (as with all rooms facing south) or around a central courtyard.

Photo: David Martindale

Many earth-sheltered designs are only partially underground and so avoid many of the difficulties of underground house design and construction. Setting the house partly into a hillside, or pushing earth up against some walls (berming), are the most common uses of earth-sheltering.

## The Superinsulated House

Parallel with the development of the various passive and active solar heating systems there has been an increase in the understanding of how heat is lost from buildings and how to minimize those losses. The culmination of efforts to improve building insulation has been the development of the superinsulated house, a building so effectively insulated and sealed against infiltration that it neither gains or loses much heat. Superinsulated houses frequently require almost no supplemental heat, even in cold climates. Most of their heating needs are met by the building's internal gains — intrinsic heat sources such as the body heat of occupants and the heat given off by cooking stoves and electrical appliances and lights — and by modest solar gain through south-facing windows.

These thermal fortresses require no thermal storage other than the normal mass of the house and its furnishings, no central heating system, and no particular orientation to the sun or arrangement of rooms. They can be planned with any conventional house shape or floor plan.

Superinsulated houses can make use of some direct gain through moderately-sized south windows. Window areas on other sides must be kept fairly small in order to prevent unacceptable heat loss — unless night window insulation is used. The degree of tightness necessary to keep infiltration losses very low usually means that an air-to-air heat exchanger must be used to bring in a good supply of fresh air, although the cost of buying and operating an exchanger is low compared to the savings which result from decreasing unwanted natural ventilation.

Superinsulated houses superbly protect their occupants from outdoor heat, cold, and noise. The methods used in superinsulated houses can be effectively used in any type of building.

**Underground House**

**Superinsulated House**

## THE ENVELOPE HOUSE
## A New Kind of Solar House

In 1978, magazine articles began introducing a distinctly new kind of passive solar house, a magical house-within-a-house that kept itself warm with an envelope of sun-heated air circulating between two well-insulated shells. This new housing concept acquired a variety of different names: it was called the "double shell solar

Published by the Whole Earth Catalog

The CoEVOLUTION Quarterly

Gregory Bate[...]
The Pattern Which [...]

No. 18 Summer 1978

# Don't Build a House till You've Looked at This

by Michael Phillips

It can be built by routine contractors, with standard technology and ordinary materials, at no extra cost, and will work at most U.S. latitudes and temperatures. Furthermore, it is completely passive — there are no fans or pumps whatsoever. And the design will work on most configurations of living space.

*Footnotes by J. Baldwin*          *Photos by Michael Phillips*

This is the lowest level of technology with the highest return that I have seen. The Kubota/ Smith house is at 7,000 feet on a mountainside overlooking Lake Tahoe. It was built by a local contractor in four months at $31.00 a square foot,[1] less than the standard local cost using standard construction techniques.

I visited the house on March 23, 1978, with Sim Van der Ryn, the California State Architect, and J. Baldwin, the **CQ** Soft Tech editor. We found it nice and warm with only solar radiation as a heat source, despite the fact that it was 8:45 in the morning, snowing outside (as you see in Photo 1), five feet of

snow around the house, and the sky had been overcast for the past three days.

The technology is very simple. The house faces south and the south face is double glazed glass. The air heated by the sun rises on the south side to the roof which has a wider than usual space (12" between roof and ceiling) where there is an opening allowing the warm air to move down through the roof rafters to the rear north wall. There it falls directly to the basement crawl space and flows under the house to rise again through the decking in the south face,[2] forming a circle. The main living area of the house is surrounded by a gently moving circle of warm air.

Why does it work? Convection. The sun warms the air during daylight hours; the air expands and rises, drawing in cooler air from under the decking and keeps the circle of air moving. This warmed air heats all the internal mass of the home until nightfall when the air slows or stops and then provides a perfect air insulation blanket to keep the warmth from radiating to the outside[3] (after midnight, the air circulates in reverse, because the south glass wall is the coolest).

1. I thought I heard him say $29.00 per square foot, which is commendably low these days.

2. The air "rises through the deck" by passing between the 1/4" gap between the deck planks. This gives a uniform circulation over a wide area rather than airflow being concentrated in a duct which would likely require a blower. The eerie silence of the house is one of its best features.

3. Experiencing the draftless silent comfort of this house makes me anxious to try some more experiments with heat radiation. I think "the book" may well have misled us all these years.          —JB

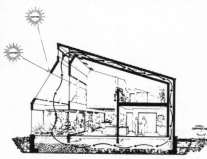

*Drawing shows path of heated air: from greenhouse, through double-ceiling, down back of the structure, across rocks in the crawl space and back into the greenhouse. Outside vent at right side draws air in the summertime.*

**100**

house," "double envelope," "buffering air envelope" and "convective loop" house. Most people came to know it simply as the "envelope" house.

The early envelope designs featured a continuous plenum, or passageway, of air surrounding most of the living space. Air heated by the sun in a south-facing greenhouse or solarium moves by convection up and through the airspace between the inner

**Better Homes and Gardens**

CIRCULATION 8,000,000
March 1979 • $1

### A Step Ahead in SOLAR LIVING

BY CHERYL SCOTT

This home has no furnace, no air conditioner, not even a fan. Yet the house maintains comfortable indoor temperatures year-round. How? Air flows in a gravity convection loop through an air space that acts as a buffer between you and the weather. Read on and we'll explain how this ingenious design works.

Lots of people think that Tom Smith's home in Lake Tahoe, California, is just another passive solar house—that is, one that doesn't use mechanical means to generate and distribute heat. The house does have a lot in common with passive solar homes. The similarities include a south-facing greenhouse in front, no windows on the north side, few windows on the east and west, plenty of insulation, no mechanical comfort equipment, and (best of all) incredibly low heating and cooling bills. But Tom Smith's house has something that other passive homes lack: a second skin.

Any house transfers heat through its roof, walls, floor, windows, and doors—all components of the building's skin. But in Tom's house, the skin that surrounds the living space is not the surface that's exposed to the weather. An insulative air space separates the "inner house" skin and acts system. Surrounding the lope of solar living areas sta in winter, coo because there's tem at work, house aren't and changes in

For more de lope system, tu

**exterior**

Construction of Smith's three-bedroom, 2,000-square-foot home called for no unconventional building materials or methods. The home cost no more to build than any house of similar square footage in the Tahoe area. The simple box shape of the house and the lack of ductwork resulted in savings that offset the additional framing costs for the extra 2x4 stud wall on the north side.

**first floor**

### SOLAR LIVING

#### How the system works

The Smith house is surrounded on four of its six sides (think of the house as a cube) by an envelope of solar-tempered air. This heat transfer envelope is a continuous space consisting of the

▶ **gaining heat**

The large areas of insulated glass in the greenhouse help to collect solar heat during the day. As the greenhouse heats up, the envelope air acts as a gravity convection loop and goes into action. (Follow the red arrows in the diagram below.) Warmed air rises in

earth-floored crawl space, the south-facing greenhouse, and a 12-inch-deep plenum that runs between the rafters along the roof and between the studs along the north wall of the house. Air flowing

the greenhouse and flows through the roof plenum. When the air reaches the cooler north wall, heat loss along the north face of the envelope allows gravity to draw the convection loop downward toward the crawl space. The mass of earth in the crawl space

within this heat-transfer envelope space can accomplish three things: gain heat, lose heat, and ventilate and cool. Follow the colored arrows in the diagram below and see how each mode works.

acts as a heat bank, absorbing the remaining warmth from the air as it flows over the earth. The airflow loop is complete when the air reaches the greenhouse. There the air again warms and rises, continuing the clockwise heat gain flow.

OUTER WALL INSULATION
1 FT. MIN. AIR SPACE
INNER WALL INSULATION

→ VENT
→ GAIN
→ LOSE

INSULATION
FROSTLINE

PRECONDITION TUBE

*The Kubota/Smith Solar House at 7,000 feet, near Lake Tahoe. Snowing at nine in the morning with 5 feet of snow on the ground. Toasty inside.*

o me the most interesting benefit of this system is at the air warmed directly by the sun is not the air ou live in. The air you live in is warmed by radiation om the house; it can be cooled by opening a window; e humidity is increased by taking a shower. The ubota/Smith's report that living in the house has en a very different experience from anywhere else ey've lived because the air is not heated by a rnace or blown around to keep it near the floor; s fresh and calm. The minimum temperature for ing in the house can be much lower than most ople are used to. It was 64° F on the first floor and asty when we visited. The Kubota/Smith's say that en 59° F is fine because all the floors and corners e warm and there are no drafts. J. Baldwin con-med this from other experimental work and his vn experience at the Integrated Living System mes in New Mexico.

he greatest benefit of this house other than the fact at it works so well, is that it uses standard construc-on techniques which are in use everywhere in the .S. Simply described, it is a well-insulated house, ith enough roof space and northwall space to allow r to circulate around inside. Although the Kubota/ mith house has rocks in the basement, Tom Smith dn't feel that they were playing much of a role. He lt that backfill (dirt removed while building a base-ent) might do just as well or better than the rocks

which were put in during winter and still hadn't warmed. The house works wonderfully. With some minor changes this house will get even warmer in future winters. Sim and J. agreed that insulation and convection played the main role and that the internal mass of the house was the prime heat storage, not the basement rocks. Both felt that rocks, if

*Sim feeling the gentle air current rising from between the floor boards.*

### Envelope Terminology

The terms "envelope" and "double envelope" have both been widely used to describe double shell designs. When "double envelope" is used, the implication is that each building shell is an "envelope." Another convention is to refer to the airspace around much of the house as the "envelope" — a thermal envelope. In this book we will usually use the term "envelope" (not double) in reference to double shell buildings, assuming that the word refers to the airspace rather than to the building skins which create that airspace.

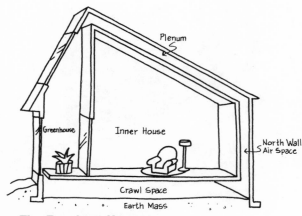

**The Envelope House**

house ceiling and the roof. Losing heat along the way, the air travels down the airspace between double north walls to a crawlspace or basement. Here, it discharges some of its remaining heat to the earth mass below the house before moving up through grates in the greenhouse floor to replace the heated air rising behind the glass.

To some extent that convective airflow reverses at night. As heat is lost through the large expanse of south glass, the other airspace surfaces and the earth release stored heat to temper the envelope air and minimize the temperature difference between the inner house and its surrounding airspaces.

Set into the spring-like climate of the envelope air, the inner house is buffered from outside temperature extremes and many infiltrating winds, and requires only modest amounts of supplemental heat to maintain a comfortable interior.

### Excitement

These houses generated widespread excitement among prospective homeowners, and occupants of the early envelope houses were unequivocally enthusiastic about their new homes. It was not difficult to understand why.

Here was a heavily insulated house permitting large areas of glass, giving plenty of daylight and a

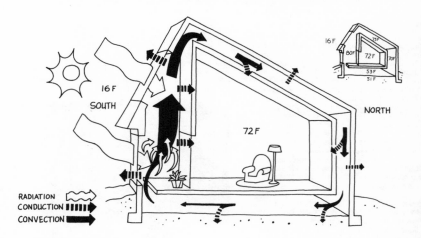

**Envelope House, Day**

On sunny days the air moves upward through the greenhouse.

wide view of the world. Concrete walls and floors, water walls, and other obtrusive forms of thermal storage were conspicuously absent; envelope house inhabitants could enjoy the resiliency and beauty of natural wood floors or soft carpeting. Here was a fully integrated solar greenhouse, visually accessible from the house and clearly a sun-drenched extension of the living space. Owners reported that these houses never became very cold, even after many cloudy days with no supplemental heat.

## Controversy

The envelope house was soon surrounded by extraordinary controversy. Some of the early proponents advanced performance claims and explanations of how these buildings worked that seemed impossible to other solar enthusiasts. Critics argued that gravity convection could not pump appreciable quantities of heat down into earth storage, and insisted that the earth is a heat sink rather than a heat source. They said that a structure's roof, typically an area of high heat loss, is hardly the appropriate location for a passage transporting large quantities of heat. Some people worried that the air circulation around the structure could make the envelope house a death trap in case of any fire in the envelope space.

### Natural Convection Loop

Natural convective airflow results from the difference in density caused by a difference in temperature from one part of a body of air to another. Heated air expands and becomes lighter, and is forced upward by heavier cooler air around it.

In a closed loop of air such as exists in an envelope house, if the air is heated near the bottom of one side, the result is a continuous flow of air around the entire loop.

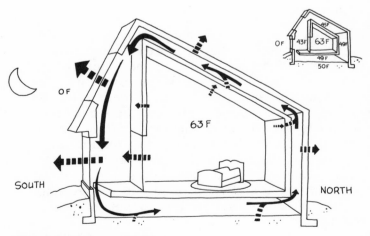

**Envelope House, Night**

The cooling from greenhouse glass makes the air movement reverse at night, with the cold air falling down in the greenhouse.

## Workshop and Survey

In 1980 an Air Envelope Workshop was presented at the Fifth Passive Solar Conference at Amherst. Many of those who attended hoped that this workshop would end the confusion and controversy.

Because we seemed still to be like blind men describing the elephant, that winter Community Builders undertook a survey to discover at least the general realities of existing buildings. In our survey of over 100 buildings, we substantiated that the houses actually were performing remarkably well (see Chapter 9). No other type of "solar" house averaged such a low heating cost.

But primarily, people questioned the cost. They felt that dividing the north wall and roof to create an airflow space had to make the cost significantly higher than the cost of a superinsulated house of similar size and thermal efficiency.

Even enthusiastic envelope house designers and builders were asking many questions of themselves and each other. What were the optimum shapes and dimensions for airflow channels? The best amount and location for south glass? How much insulation and where? How best to deal with the high humidity of a tightly-built, earth-connected structure? Could costs be reduced without sacrificing the thermal efficiency or the other exciting benefits the house provided?

Several technical conferences held during 1979 and 1980 allowed designers and builders to share their experience with double shell structures. Some envelope houses were instrumented for monitoring, although they were not necessarily representative. As monitoring data began to accumulate, it became evident that these structures were thermodynamically more complicated than early designers and critics had imagined. There were so many intricate and subtle interrelationships among the various design elements that it was much more difficult to analyze these houses than the more straightforward superinsulated or direct gain houses.

## Evolution

Although there has been much copying of early designs, people also began to experiment with design modifications, and envelope concepts began to broaden. Some designers incorporated more of the living space or other useful space into the air envelope. Some severely reduced the size of the airflow spaces. Some retained a buffering airspace only at windows, where heat loss would be the greatest. Another design variation retained the earth-connected greenhouse and used a small fan to pull high-grade heat from the top of the greenhouse into a heat storage system built under the floor of the house.

Many of these design modifications appreciably reduced the cost of full double-shell construction without affecting the superb thermal performance

After ten years' intense development of solar and energy-conserving housing, certain methods and approaches stand out for combining high performance with economy and satisfaction—

Building for Energy Independence

# Sun/Earth Buffering and Superinsulation

224 pages
—285 illustrations

Don Booth with
Jonathan Booth and
Peg Boyles

TYPICAL
PROPORTIONS
OF
HEAT LOSS

ROOF 11%

INFILTRATION 30%

WINDOWS 22%

WALLS 24%

13% FLOOR

An award-winning solar pioneer, long known for building quality homes, shares professional, state-of-the-art information in a clear, concise, and highly readable book—a practical, fully-illustrated guide to understanding, designing, and building the new energy-independent homes that are a joy to own.

"**AN ALL-AROUND FIRST-RATE BOOK** on the newest and best options in energy-efficient housing design. Packed with practical advice based on the author's 30 years' experience in anticipating difficulties and achieving trouble-free, low-cost construction.

"I am continually amazed at how much very specific advice, including ideas new to me, is included. Not vague generalities and repetition of well-known advice, but specific and fresh new advice."
—*William Shurcliff*

**BASICS OF ENERGY INDEPENDENT DESIGN AND CONSTRUCTION**—Principles, guidelines, and systems applicable to any energy-efficient house design, explained and illustrated in detail. Covers the practical information you need, such as:
- How to install a continuous vapor barrier
- Using new products to control air infiltration
- Ten cost-effective, high R-value wall systems
- How to control condensation and prevent moisture damage
- Siting and solar access
- Cost analysis of various approaches
- Fire safety considerations
- Small heating systems for efficient houses

**PRINCIPLES & COMPARISONS** — Guidance for choices
- How the new energy-conserving homes work: types of solar design, earth sheltering, superinsulation
- Comparisons of advantages and limitations
- Making the best choice in individual situations

**SUPERINSULATION**—The biggest new trend in energy-efficient housing is the simple, direct approach: keep the heat loss from a house so low, by using thick insulation and controlling air leaks, that very little heat is needed to keep the house comfortably warm, even in the coldest weather.
- How to use solar heat without added thermal mass
- How to plan for ample daylight in a super-insulated house
- How to provide fresh air in an airtight house
- Heavy insulation vs. superinsulation: how to decide how much
- Examples of superinsulated houses, with plans and photos

**SUN/EARTH BUFFERING**—The "envelope" house stunned people with its tiny fuel needs and charmed them with its sunlit greenhouse/solarium, but the concepts were controversial and many people balked at the expense of double wall construction. Newer "buffered" designs can retain the advantages while simplifying construction and reducing costs. The earth-tempered sunspace is still a vital feature of the buffered house, warming the house while providing a sun-filled living space which can be used for year-round food production.
- How to keep the advantages of the "envelope" without the double walls
- How to provide large glass areas without high heat loss
- How to use the earth's heat to protect a greenhouse from freezing
- How to design a greenhouse for good plant growth
- Tips on greenhouse management
- How to distribute sun-heated air, with or without fans
- Report on a survey of over 100 envelope homes
- Twenty examples of buffered houses from across the continent

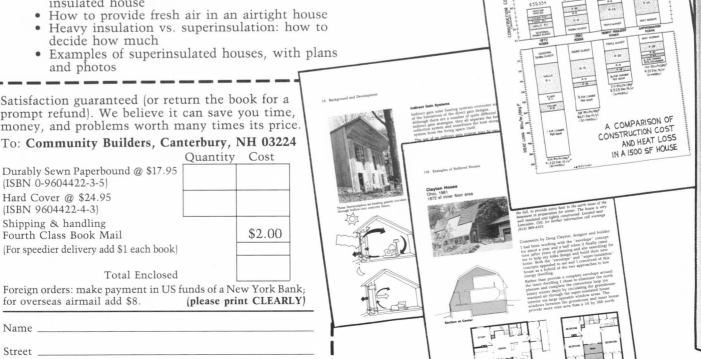

A COMPARISON OF CONSTRUCTION COST AND HEAT LOSS IN A 1500 SF HOUSE

**FOR THE ARCHITECT:** A practical guide to energy-efficient house design with many detailed drawings. Emphasizes systems that give superior performance at reasonable cost. Includes a concise survey of the whole field, explaining the pros and cons of each approach. Provides an abundant source of tested design ideas.

**FOR THE BUILDER:** Brings you up to date with detailed, illustrated information on new construction methods for energy-saving homes, with the reasons for new practices clearly explained. Helps you avoid costly mistakes.

**FOR EVERYONE WHO WANTS TO BE WELL INFORMED:** A clear explanation of the principles of energy-conserving housing. Gives a fascinating look at a wide variety of successful homes, with photos, floor plans, and commentaries. Supplies knowledge required for wise design choices and for informed awareness of good construction practices. INDISPENSABLE FOR ANYONE THINKING OF BUILDING.

**"RECOMMENDED FOR THE NOVICE AS WELL AS THE EXPERT.** Rarely are we blessed with as unique a combination of raw field experience, writing skill, and dedication to ideals as Don Booth brings in this book."
—*Bruce Anderson*

Take advantage of the best information now available—

The house you build now will be almost new in 1990 and will get most of its use after the year 2000. As fossil fuels grow scarcer and prices go higher, many homeowners will be saying "I wish I had." This book was written to give you the chance to say "I'm glad I did."

**"I HAVE JUST HAD THE WONDERFUL EXPERIENCE** of spending a delightful three hours with SUN/EARTH BUFFERING and SUPERINSULATION. This will be a tremendous help. We have needed such an authoritative book on this subject."—*Howard Faulkner*

---

**Cross-Hatch Details**

- roof plywood
- airspace for venting
- rafters w/R-30 fiberglass
- crossed nailers w/R-13
- Tyvek
- wall frame w/2xR-13
- diagonal bracing
- vapor barrier
- header and joists
- foam insulation
- gypsum wallboard
- operable window unit
- fixed glass

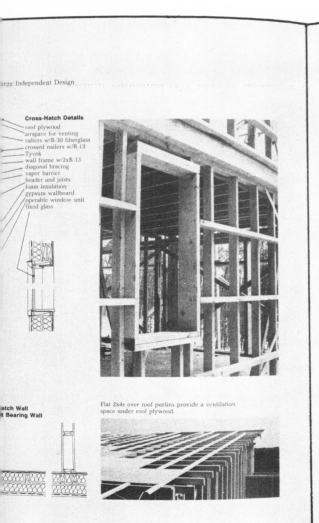

Flat 2x4s over roof purlins provide a ventilation space under roof plywood.

atch Wall
t Bearing Wall

The OUTRIGGER starts with a standard plywood-sheathed frame wall and can be used for retrofit superinsulation of an existing house. Vapor barrier sheeting is applied to the wall exterior, and preassembled outrigger frames are nailed to the walls. The frames as used by Gerald Leclair in Winnipeg, Saskatchewan, consist of vertical 2x4s with plywood gussets and nailer blocks for connecting at floors and roof. (A "Larsen Truss" by John Larsen of Edmonton, Alberta, uses two 2x2s with gussets every two feet. Addresses for purchasing additional information on both systems are in Reference Section D on page 211.) Cost here includes the standard sheathed wall.

**#10 — 8" Outrigger on 2x4 wall**

|  |  |  |  |
|---|---|---|---|
| Typical cost | $2.84 | ($3.13) | ($3.07) |
| R-value: | 31 | (44) | (39) |

**HUMIDITY AND VAPOR BARRIERS**

Loose houses with a high rate of infiltration tend to be very dry in the winter, since there is a constant influx of cold outside air and when this cold air is heated, its relative humidity becomes very low. Well-insulated tight houses do not have this large source of cold dry air, so the moisture created by the natural occupancy of the house tends to build up, creating higher levels of relative humidity.

This can be a more healthful atmosphere for people, but in cold climates the exterior building shell stands between warm moist air and cold dry air, and the shell must be designed and constructed with great care.

Warm humid air in a house cools as it moves out of the structure toward the cold outside, and at some point in that cooling process it will reach the dew point (100 percent relative humidity) and vapor in it will condense. In an insulated wall, this dew point will typically be reached somewhere within the insulation. The resultant build-up of water can cause serious problems, particularly the deterioration of insulation and the destruction of paint and wood.

**9— 12" "Saskatchewan" unit wall**

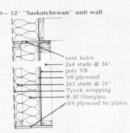

- vent holes
- 2x4 studs @ 24"
- poly VB
- 3/8 plywood
- 2x3 studs @ 24"
- Tyvek wrapping
- R-30 fiberglass
- 3/4 plywood tie plates

**10— Outrigger on standard wall**

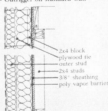

- 2x4 block
- plywood tie
- outer stud
- 2x4 studs
- 3/8" sheathing
- poly vapor barrier

**Relative Humidity and Dew Point**

The amount of water vapor that a volume of air can hold is dependent on the air temperature. The RELATIVE HUMIDITY (RH) of air is the percentage of water contained in the air compared to the maximum amount that the air could hold at its particular temperature. Warm air can hold much more water vapor than cold air can hold. The relative humidity of a volume of air containing a particular amount of water decreases as the air is heated, since the maximum amount that the air can hold is increased. Conversely, the relative humidity increases as air is cooled.

The DEW POINT is the temperature at which air is cold enough to have a relative humidity of 100 percent. When further cooled, some of the water vapor condenses to form liquid water.

# CONTENTS

**Community Builders**
**Canterbury, NH 03224**

Address Correction Requested

PREPAID PRICE FOR ADDITIONAL COPIES $12.95 (SOFT), $19.95 (HARD) PLUS $2.00 FOR SHIPPING.

Before you plan a new house—

Are you aware that it is now practical to build homes that heat with less than one cord of wood or $200 worth of electricity yearly—even in northern New England?

—Not just homes on south slopes or underground, but comfortable homes built *on any site* and *in any style.*

**An important new book tells you how . . .**

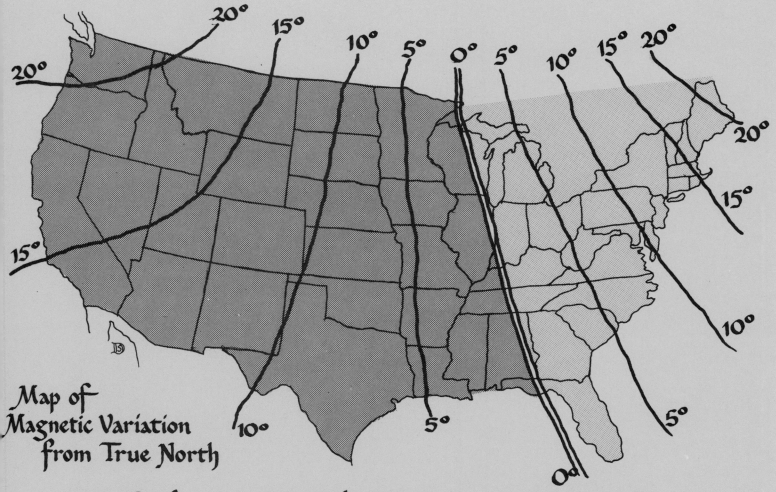

Map of
Magnetic Variation
from True North

# ~ Solar South Locator Card ~

## First Method

One of the first steps in siting a house is to determine which direction is true south. The easiest way to do this can be with a compass.

Unfortunately, a compass points to magnetic north, which in most locations is not the same as true north. The amount of this difference varies across the US, from about 20 degrees in one direction to about 20 degrees in the other direction. Variations across the US are shown on the map above. If you happen to live on the zero-degree line that runs through Wisconsin and Illinois, you don't need this card, because compass south and true south are the same.

This card describes two methods for locating solar south with your compass; using both methods can provide a double-check. The first method requires a compass with degree lines marked on it.

1. Find the variation line closest to your location on the map above. Note the number of degrees and whether it is to the <u>left</u> (on the dark side) or to the <u>right</u> (on the light side) of the zero-degree line.

2. Line up with your compass so you are facing compass south.

3. If your location is shown on the LEFT (dark) side of the map's zero-degree line, turn to your LEFT the correct number of degrees to find true solar south. If your location is shown on the RIGHT (light) side of the map, turn to your RIGHT the correct number of degrees.

(see other side of card)

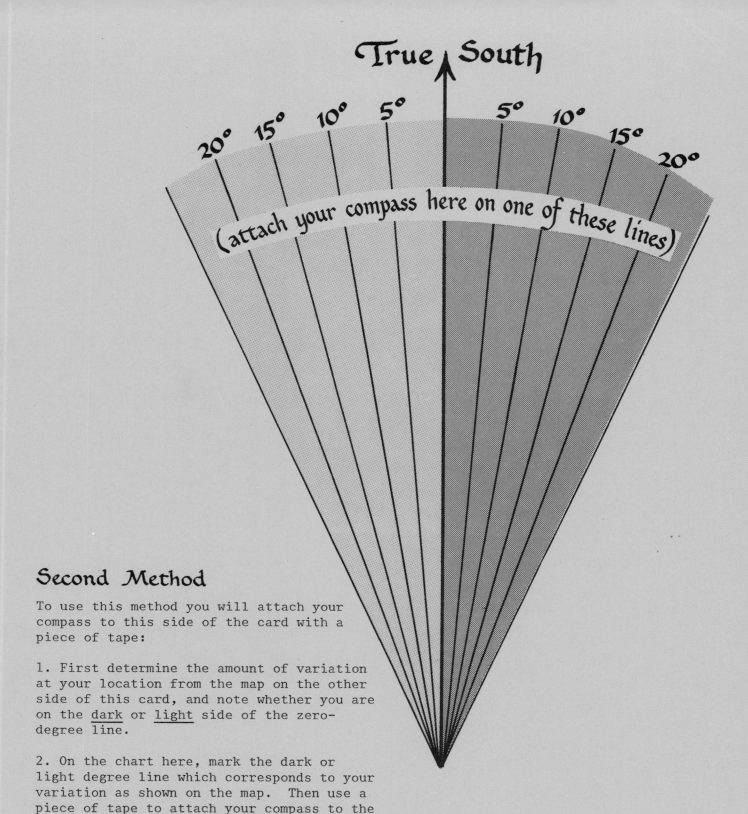

True South

20° 15° 10° 5° 5° 10° 15° 20°

(attach your compass here on one of these lines)

## Second Method

To use this method you will attach your compass to this side of the card with a piece of tape:

1. First determine the amount of variation at your location from the map on the other side of this card, and note whether you are on the dark or light side of the zero-degree line.

2. On the chart here, mark the dark or light degree line which corresponds to your variation as shown on the map. Then use a piece of tape to attach your compass to the card so the north-south centerline of the compass body lines up directly over that marked line (don't worry yet about which way the needle is pointing).

3. To take a reading: Turn the card with the compass so that the compass needle lines up with the compass body centerline. The south arrow on the card now points to true solar south.

or eliminating any of the other advantages people found so exciting in envelope houses: the large greenhouse which could be opened to the living space, a pleasant interior without glare, and protection from freezing.

Lacking a better name, most designers were calling even their innovative variations "envelope" houses, although these buildings no longer fit the early definitions of a double shell house. A new name was needed, a broad generic term that would embrace all of these related house designs, from the double-shell prototypes to the newer designs, and including future buildings not yet conceived.

Because all of these buildings are uniquely buffered by sun-heated airspaces connected to the earth itself, we suggest the name SUN/EARTH BUFFERED HOUSE. The next chapter introduces the buffered house as a radical new building system, and discusses just why these houses work so well and, with all their complexities, accomplish so much so simply.

Amherst workshop speakers, left to right: Standing: Robert Henninge, Homer Ghaffari, Jim Berk, Ken Ortega, Rick Cowlishaw, Jim Ray, Bill Brodhead, Bill Shurcliff, Vic Reno, Nick Nicholson, Doug Holmes, Bruce Maeda, Tom Smith, Bing Chen. Seated: Robert Smith, Don Dougald, Gerald Kitzman, James Akridge, Ralph Jones, Norman Saunders. On floor: Hank Huber, Don Booth, Bill Mead.

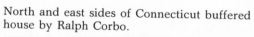

Condominiums at Oakland, CA, by Lee Porter
Butler, Ekose'a.

(In center) Greenhouse apartments at Grand
Junction, Colorado, by Malcolm Lillywhite,
Solstice Designs.

North and east sides of Connecticut buffered
house by Ralph Corbo.

Hooker Chemical Center at Niagara Falls has double exterior walls four feet apart, with operable louvers that make each side of the building a ''living skin'' adapting to weather changes. Architects: Cannon Design.

Michigan buffered house by Paul R Bilgen.

# 3

## HOW THE BUFFERED HOUSE WORKS

### Familiar Systems Uniquely Combined

Buffered or envelope houses, for all their novelty, contain no single new functional element not already a familiar feature of other energy-efficient building systems. Attached solar greenhouses, heavily insulated walls, the use of earth heat for tempering part of a home's interior, and natural convective airflow for moving heat around inside a building are all strategies which have been employed successfully in other building designs.

What made the envelope houses such a radical departure from earlier shelter systems, and what explains the spectacular thermal performance of all well-designed buffered buildings, is a unique combination of several functional design elements, individually familiar, but never before allied in quite the same way. It is this combination of functional elements that distinguishes a sun/earth buffered building.

We have chosen to recognize six functional design elements in these buffered structures: insulation, the buffering airspace, the earth connection, solar gain, convection, and thermal storage. Though these six include the essentials, the list is somewhat arbitrary, and it might be argued, for example, that the earth connection is actually a kind of thermal storage, or that the specialized function of the greenhouse demands recognition separately from other buffering airspaces.

Determining precisely how many elements distinguish buffered buildings from others is not crucial to understanding these structures. More important is to recognize that the various essential elements of a well-designed buffered building do not function independently, but work together in one elegant system that offers substantially more advantages to a homeowner than previous design concepts.

## THE ESSENTIAL ELEMENTS

**1. Insulation:** The first essential element of any energy-efficient housing design is effective insulation and control of infiltration. Superinsulated houses have demonstrated that by reducing the heat loss from a structure sufficiently, the auxiliary heat required to keep it warm even in a very cold climate can be reduced to a negligible amount. Envelope and other buffered houses employ many of the same strategies.

Designers of superinsulated houses must either use fairly moderate glazing areas or else use night window insulation to prevent unacceptable nighttime and cloudy day heat losses. In contrast, buffered houses can contain large glazed areas and provide

### The Effect of a Buffering Airspace on Heat Loss

In the first drawing below, the interior of the house (on the left) loses heat just to the buffering airspace between the double window and wall. The airspace is assumed to be at 50F. In the second example, the total R-values of the double window and wall have been combined, but the interior is losing heat directly to the 10F outside. Because of the increased temperature difference in the second example, the heat loss is significantly greater.

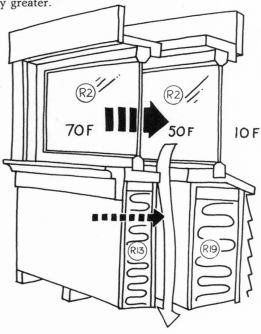

$$\text{WINDOW HEAT LOSS} = \tfrac{1}{2} \times 20 = 10 \text{ BTU/HR}$$

$$\text{WALL HEAT LOSS} = \tfrac{1}{13} \times 20 = 1.5 \text{ BTU/HR}$$

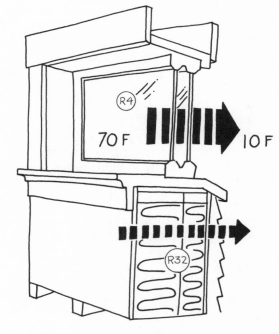

$$\text{WINDOW HEAT LOSS} = \tfrac{1}{4} \times 60 = 15 \text{ BTU/HR}$$

$$\text{WALL HEAT LOSS} = \tfrac{1}{32} \times 60 = 2 \text{ BTU/HR}$$

cheery daylit interiors and visually accessible greenhouses without excessive heat loss or overheating, because the largest glazed areas from the interior are placed to lose heat only to the tempered climate of a buffering airspace.

**2. Buffering Airspace:** Since the amount of heat lost through a section of material is proportional to the temperature difference between inside and outside, exposing the vulnerable parts of the heated house to a moderately warm airspace, instead of directly to cold outside air, reduces heat losses significantly.

This is the function of a buffering airspace. It is a space of moderate temperature between part of the inner heated house and the outside, thermally separated from each by a window or insulated wall.

The effect of a buffering airspace is most pronounced at windows, with their characteristically low resistance to heat loss, and in most buffered houses the largest windows overlook the greenhouse. So the greenhouse itself is the most important buffering airspace, and maintaining a moderate temperature there both protects the sunspace itself from getting too cold and reduces heat loss from the inner living space. While contributing much to the superior thermal performance of the whole building, the greenhouse also adds an important multi-purpose space to the house, providing an area for four-season food production, and a sunny, glassed-in space for sitting, eating, or sunbathing (although without tanning).

The total amount of buffering airspace which is used in addition to the greenhouse/sunspace varies in the many different design possibilities. A double-shell envelope house has an airspace above the whole ceiling as well as between two complete north walls. One of the important criticisms of the early envelope houses was that creating full north wall and above-ceiling airspaces added substantially to the cost of the house without adding useful space or even contributing much to thermal performance. In many newer buffered houses, designers have reduced or eliminated north wall and above-ceiling airspaces, finding it more economical to use heavier insulation to keep heat loss low through those areas, and using smaller airshafts for taking heat into storage.

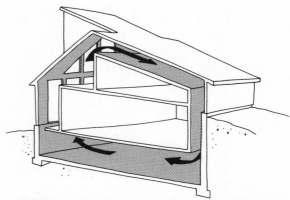

**Buffering Airspaces in a Typical Envelope House**

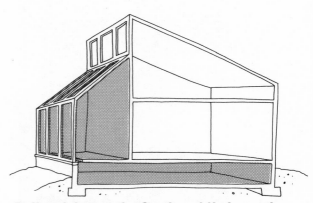

**Buffered Just at the South and Underneath**

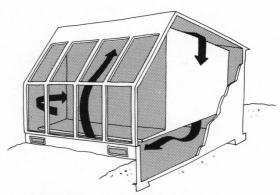

**Buffering Airspaces on All Sides**

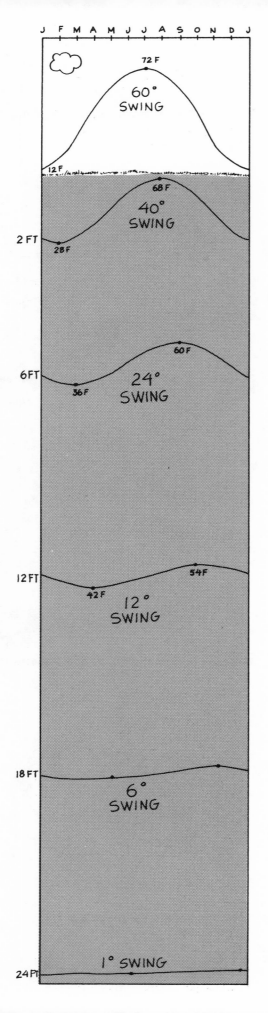

Seasonal temperature swings are smaller at greater earth depths.

The value of a buffering airspace in reducing heat loss, especially through large glazing areas during very cold weather periods, is apparent. The question of how the airspaces can be kept relatively warm even during sunless winter periods is answered in large part by the role of the third functional design element, the earth connection.

**3. Earth Connection:** Virtually all houses have a thermal connection to the mass of the earth. What is unique to buffered houses is that the earth connection is made to all the buffering airspaces. rather than just to an enclosed basement or directly to the house itself through a slab floor.

The earth beneath a house is an enormous mass that contains a tremendous amount of low-grade heat. It is not warm enough to keep a house comfortable in a cold climate — around 50F is too cold for comfort — but it can be very effective in moderating the temperatures of buffering airspaces as they gain heat on a sunny day or lose heat at night. This is the primary function of the earth connection in a buffered house: keeping the buffering airspaces, particularly the greenhouse, from getting too cold or too hot.

The temperature of deep-down earth is nearly constant year-round, a few degrees above the annual average air temperature at a given location. This means that in winter the deep earth will be substantially warmer than the outside average temperature. The great mass of the deep earth stabilizes the temperatures of the near-surface earth or concrete. With some convection, the floor of the basement or crawlspace either absorbs or releases heat which moderates the temperatures of all the connected buffering airspaces. The earth connection is this free flow of heat between the earth and the buffering airspaces.

One of the early criticisms launched at the envelope design was that the earth was an inefficient heat storage medium, and that it represents a long-term continual heat loss. Most buffered house designers do not now see the earth as an efficient form of daily heat storage, but rather as a great thermal flywheel, a beautifully simple, inexpensive and reliable way to moderate the buffering airspace temperatures during extremes of outside weather.

**4. Solar Gain:** In a buffered house, the greenhouse or sunspace receives most of the direct solar gain. Sunshine provides heat and light there for people and plants, and when that space starts getting hot the excess heat is moved by convection to storage elsewhere. Although the interior living space usually receives relatively little direct solar gain, it can still receive heat and diffuse daylight through large windows and glass doors overlooking the sunspace.

One of the basic problems that any solar designer must address is the fact that increasing solar gain with larger window areas causes a corresponding increase in heat loss, since windows are such poor insulators.

This problem can be very pronounced in attached greenhouses. In order to maintain acceptable plant-growing temperatures without auxiliary heating during sunless periods, solar greenhouse designers have relied on massive heat storage systems that store heat on sunny days for later release back into the greenhouse. But the more heat stored for later use in the greenhouse, the less heat is available for the rest of the house, causing many designers to choose between maximum food production or maximum heat for the house. Balancing these concerns can be difficult.

In a buffered house, the earth connection — with some periodic recharging — can supply enough low-grade heat to keep the sunspace moderately warm (say 45-50F) even during sunless midwinter periods. This allows the designer a variety of choices for the storage and use of much of the heat collected by the greenhouse on sunny days.

**5: Convection.** There are various methods used for transporting heat into and out of storage in a buffered house, but they all use some form of the fifth functional design element, convective airflow. Heated air is transported through the buffering airspaces by either *natural convection*, caused by the effect of gravity on air at different temperatures and densities, or *forced convection*, in which fans or blowers are used to move the air to desired locations.

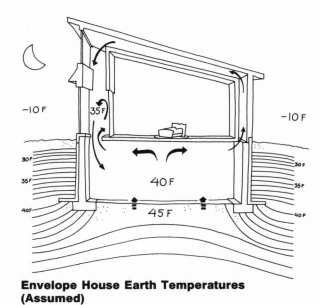

**Envelope House Earth Temperatures (Assumed)**

• *Natural convection:* In the original envelope designs, air was observed to circulate by natural convection in a loop encircling the inner house. In most cases the airflow rates in those houses was highest during the daytime, when the air moved upward through the sunspace, and less pronounced at night when it moved in the opposite direction.

One of the controversies surrounding the envelope houses was the question of whether or not natural convection alone could reliably move enough air and heat down into the basement or crawlspace. Temperatures recorded in monitored buildings showed that some envelope houses were clearly effective in moving heat by natural convection from the greenhouse into storage, although much of that storage was in the materials of the house itself and not in the earth. In other cases the sunspace was overheating, indicating insufficient natural convective transfer of heat.

The airflow rates achieved by natural convection apparently depend on a variety of factors. Low south glazing tends to move more sunspace air upward, giving a better push to the entire loop. Greenhouse floors of spaced sun-bathed planking may improve a convective flow by their spread-out transfer of heat to the air. Surface friction and turbulence at constricted airflow spaces tend to slow the volume of flow.

• *Forced convection:* Fans or blowers can move much larger quantities of heat directly to where it is wanted. A fan can quickly take high-grade heat from the top of the sunspace to basement or other storage.

The use of a fan allows smaller airflow passages while still ensuring the transfer of heat to the desired locations. It means that a designer does not have to worry about the complexities of natural convection and can be confident that sunspace heat will be moved into storage. People who want large amounts of high greenhouse glass may need a fan to speed the removal of heated air into storage. The use of a fan also means that heat can be moved to locations which would be inaccessible with natural convection, such as through very constricted passages.

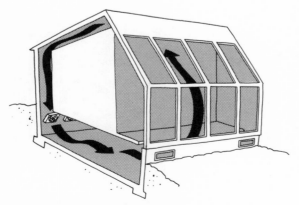

**Envelope House with Fans**

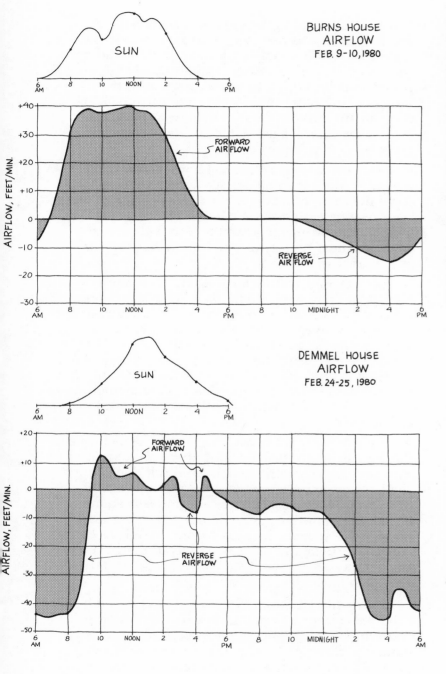

BURNS HOUSE
AIRFLOW
FEB. 9-10, 1980

SUN

FORWARD
AIR FLOW

REVERSE
AIR FLOW

DEMMEL HOUSE
AIRFLOW
FEB. 24-25, 1980

SUN

FORWARD
AIR FLOW

REVERSE
AIR FLOW

## Burns and Demmel Houses Airflow Graphs

These two graphs illustrate some of the differences between natural convective airflow patterns that can occur in envelope houses. Both houses are fairly standard envelope houses (see Chapter 8 for plans and photos). The airflow is shown in linear feet per minute, as measured in the north wall airspace. Forward airflow refers to air movement up in the sunspace and down in the north wall cavity. The cross-sectional area of the north wall cavity in both houses is about 20sf.

In the Burns house, the forward daytime airflow is more than twice as fast as the reverse flow at night. This is similar to results obtained at two other monitored envelope houses (the Tom Smith House and the Omaha Passive Solar Test Facility). In the Demmel house the opposite situation occurs: the reverse airflow is more than twice as fast as the forward daytime flow.

(Demmel house data courtesy Bing Chen.)

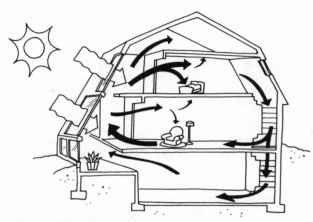

**Clayton House Heat Storage**

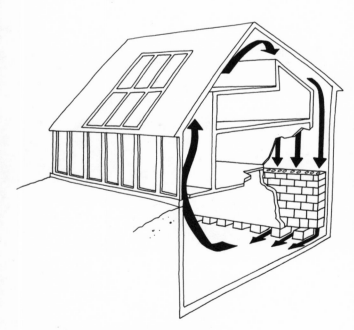

**Torii House Heat Storage**

This house has an extra concrete block wall between the two north walls, for added heat storage. Fourteen small fans blow warm air down through and around the concrete blocks on sunny days.

The cost of operating a small fan occasionally is low (less than $50/year) when compared to the life-cycle cost of full double-shell construction, and in fact the energy used to run it is eventually converted to heat. With a direct connection to the sunspace, the buffering effect of the earth connection can still protect these houses from freezing, even when a prolonged winter power outage prevents fan operation.

**6. Thermal Storage:** Buffered houses may contain several distinct thermal storage systems. These serve the same function as the thermal mass in any solar heating system: as a damper on the temperature swings that would otherwise occur when the heat source, the sun, alternately appears and disappears.

— The materials and furnishings of the home's interior store heat on sunny days. Doors and windows to the sunspace are opened, and heated air moves by convection into the house, raising the inside temperature and delivering heat to building materials and furnishings. This type of storage is of course greatly reduced if no one is at home during the day to open windows or doors, unless other means such as automatic blowers or door openers are provided.

— In many buffered houses, the materials that line the airflow spaces provide an important thermal storage mass. The airflow spaces in full double-shell houses are often lined with gypsum for fire protection, which contributes a significant amount of mass. Monitoring results indicate that in these houses, the airspace materials store more heat on a daily basis than any other materials in the structure. In houses with reduced airflow spaces, this type of storage plays a smaller role.

— In virtually all buffered houses, some of the heat collected in the sunspace makes its way into the earth or concrete beneath the house and provides a moderate amount of daily heat storage for night-time release. Recharging the surface of the earth mass also slows the gradual discharge of its deeper-earth heat during the coldest months.

— It is possible to use fans to move higher-temperature heat into an additional storage system in an unobtrusive location such as under the floor of the living space (see the Julian and Shrewsbury houses in Chapter 8).

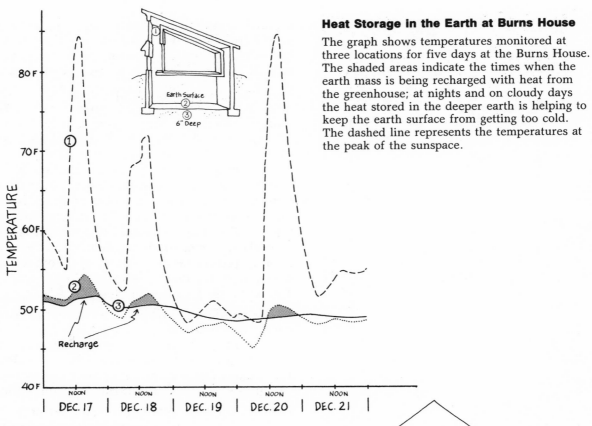

## Heat Storage in the Earth at Burns House

The graph shows temperatures monitored at three locations for five days at the Burns House. The shaded areas indicate the times when the earth mass is being recharged with heat from the greenhouse; at nights and on cloudy days the heat stored in the deeper earth is helping to keep the earth surface from getting too cold. The dashed line represents the temperatures at the peak of the sunspace.

## BENEFITS OF BUFFERED HOUSES

A wide variety of specific designs can result from an understanding of the functions and interaction of these elements of buffered buildings. Other than demanding solar access for the south-facing sunspace, buffered house design places no intrinsic limitations on floor plan, shape or size. The incorporation of a sunspace in the design creates exciting architectural and aesthetic opportunities.

Buffered houses can be nearly as low in initial cost as any building of comparable size and quality. The life-cycle cost can be lower for those who take advantage of the possibilities for food production or passive hot water pre-heating in the greenhouse.

These houses can be designed to be almost maintenance-free, placing no demands on their occupants other than perhaps opening vents at the top of the sunspace each spring and closing them in the fall. The daily winter discipline of putting night insulation into place is unnecessary. The maintenance associated with active solar heating

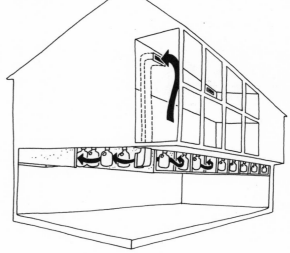

## Julian House Heat Storage

The Julian House uses an added heat storage system consisting of 850 one-gallon jugs filled with water underneath the floor. Two fans blow heated greenhouse air down through a duct and around the jugs which hang between the floor trusses.

systems is avoided. Winter vacations are relatively worry-free: even without any auxiliary heating, household plumbing in a properly designed and constructed buffered house will not freeze.

In terms of overall thermal performance, these houses are outstanding. In our survey of envelope houses which had been through at least one winter (see Chapter 9) more than half used less than one Btu/Degree-day/sf of auxiliary heat. This compares with averages of about 8 Btu/DD/sf for a survey of Northeast solar homes, and about 4 Btu/DD/sf for houses described in the Housing and Urban Development/Department of Energy (HUD/DOE) book, "First Passive Solar Home Awards." (One Btu/DD/sf corresponds to a cold-climate annual heating requirement of about one cord of wood, or about $250 for electricity at $.07/kWhr, for a 1500sf house.)

In the summer, buffered buildings stay naturally cool, protected from direct sunlight and glare by the greenhouse, and from overheating by the thermal sink of the earth mass beneath the house. Heat rising behind the greenhouse glass can be readily exhausted by natural convective airflow.

Buffered houses, like superinsulated and underground houses, protect not only against the extremes of outside weather, but also against unwanted outside noise, creating a peaceful and quiet indoor environment. Yet for all their thermal efficiency and acoustical privacy, these houses are not isolating fortresses; they also allow their occupants to maintain important visual connections to the world outside.

The sunspace integrates the house with the natural world. On sunny winter days owners delight in their bright, airy, plant-filled sunspace rooms.

The combination of benefits provided in buffered houses gives occupants a most important set of tools for resisting some of the effects of detrimental economic and environmental changes in the modern world. These tools are both physical and psychological, and the importance of each is in-increasing all the time. Part of the beauty of the buffered housing system is that it provides all these benefits intrinsically, as part of the basic shelter itself.

# 4

## BASICS OF ENERGY
## INDEPENDENT DESIGN

The design section of this book is divided into
three separate chapters. This chapter includes
design considerations applicable to virtually any
type of energy-efficient house, including buffered
and superinsulated houses. Most of the techniques
of superinsulation are covered in this chapter.
Specific considerations of buffered and superinsulated house design are covered in Chapters
5 (buffered) and 6 (superinsulated).

### Our Background

Any approach to house design is based in large part
on the designer's particular experience, judgment,
and taste. At Community Builders, our background
for design includes thirty-five years of actually
building individual houses that were intended to be
free of maintenance problems and moderate in
cost, as well as comfortable, convenient, and
aesthetically satisfying. Working with the owners of
these houses and living near enough to them to
receive their feedback has strengthened those goals.

Working as carpenters as well as designer-builders,
much of our immediate focus has been on learning
the best way to do everything that might need to
be done in the building of a house, from the
specific trade skills of sawing straight and "leaving
half the line" to buying the most durable paint. We
enjoy looking for better ways to do everything;
innovation is normal, and we have no final best
ways, for conditions will change and our
awarenesses will be different.

Sunspace by Robert Heaton.

We are writing this book to share what we have learned with designers, builders, and prospective owners, intending to make it easy for them to apply the information in their own creative ways. Readers familiar with some of the subjects covered are expected to skip over parts that seem too elementary.

## BEGINNING TO PLAN
### First Considerations: Needs and Resources

The design of each trouble-free, long-lasting, life-enhancing and economical building combines commonly known elements and systems with fresh approaches appropriate to individual circumstances.

People have their own particular needs, and it is important to identify and be sensitive to these. But needs change, too, and it helps also to try to see projects and plans from other perspectives than the present. How might present needs change in ten years? Twenty, or fifty years? How would future occupants of this house feel about each element? Does the design have enough general appeal that it would sell at an acceptable price? A house is built not just for the present or the next few years but for many years ahead.

Early design discussions should include the desired final living conditions in the house: what are the desired comfort conditions and temperatures in each part of the house at different times? Will residents be home when the sun shines, to open doors or windows onto a greenhouse, or take out window insulation?

Durability is important, and easy maintenance, but the reduction of long-term maintenance costs usually requires somewhat more initial spending. To pay for the immediate expenditures that are necessary to achieve long-term savings may require deferring some other intended expenditures. Making wise choices can be difficult, but the general approach is to include at the beginning those elements which would cost more if they were added later, and defer some elements that could be easily and economically added. Interior finishes are commonly deferred, for example.

## Owners' Financial Planning

Before actual design work starts, prospective owners should always take time for careful financial planning. Discuss with a bank the costs of borrowing. Find out the size of monthly mortgage payments the bank would consider feasible for you and how large a loan you can therefore make. Too often, carefully thought-out designs have had to be shelved or extensively revised after a trip to the bank to arrange for a construction loan. It is far better to be clear from the start about the cost limitations that will have to be respected.

As design planning proceeds it is a challenging task to keep the plans within the cost limits. In order to prevent getting too far down a track that will prove to be overly expensive, it may be necessary to have estimates done at more than one stage of the planning. This costs money too, but may prevent large revisions later. Pricing plans before they are complete has hazards to be aware of; for example, the owner may be envisioning much more expensive finish work than the estimator has assumed in the preliminary pricing.

The cost of a septic system is a large expense that cannot be much controlled but can be determined by investigation. Septic system costs can be so high as to be prohibitive. An experienced septic system designer can estimate the cost in a particular location; digging a test pit may be a necessary preliminary. The cost of a septic system should be considered in connection with the cost of the land; a low-priced lot is not a bargain if it requires an $8,000 septic system.

The chart on the next page shows the proportionate costs of our breakdown of categories in the construction of some of our houses. These reflect our experience with these particular houses and the conditions at the time they were built. Without detailed information about what was included, the chart is only generally indicative of the proportionate costs of the major elements of a house.

| CATEGORY | HOUSE A COST | % | HOUSE B COST | % | HOUSE C COST | % | HOUSE D COST | % | HOUSE E COST | % |
|---|---|---|---|---|---|---|---|---|---|---|
| Site | 10670 | 15 | 16620 | 21 | 8627 | 14 | 10977 | 14 | 6021 | 11 |
| water | 620 | | 2019 | | 3143 | | 2411 | | 1732 | |
| septic | 4600 | | 3000 | | 3300 | | 5200 | | 1500 | |
| Concrete | 5205 | 7 | 2486 | 3 | 3000 | 5 | 3354 | 4 | 1155 | 2 |
| Masonry | 1610 | 2 | 4405 | 6 | 1327 | 2 | 902 | 1 | 0 | 0 |
| Framing | 13640 | 19 | 11003 | 14 | 13974 | 22 | 13899 | 18 | 11216 | 20 |
| Roofing | 1807 | 3 | 1204 | 2 | 1286 | 2 | 792 | 1 | 1551 | 3 |
| Windows | 5060 | 7 | 4296 | 6 | 5705 | 9 | 8877 | 12 | 4053 | 7 |
| Doors | 3300 | 5 | 2249 | 3 | 2186 | 3 | 2080 | 3 | 3697 | 6 |
| Insulation | 3740 | 5 | 1942 | 2 | 2800 | 4 | 3000 | 4 | 2700 | 5 |
| Ext. Finish | 3410 | 5 | 3612 | 5 | 3467 | 5 | 4457 | 6 | 5500 | 10 |
| Gypsum | 2640 | 4 | 5018 | 6 | 3766 | 6 | 4492 | 6 | 3608 | 6 |
| Cabinets | 3080 | 4 | 2004 | 3 | 2500 | 4 | 3461 | 5 | 2666 | 5 |
| Floors | 2112 | 3 | 1758 | 2 | 3000 | 5 | 3600 | 5 | 2200 | 4 |
| In. Finish | 2200 | 3 | 2147 | 3 | 2700 | 4 | 3100 | 4 | 3450 | 6 |
| Ceramic | 132 | 0 | 488 | 1 | 0 | 0 | 0 | 0 | 0 | 0 |
| Ext. Paint | 440 | 1 | 930 | 1 | 600 | 1 | 600 | 1 | 800 | 1 |
| In. Paint | 550 | 1 | 3004 | 4 | 1800 | 3 | 2000 | 3 | 1600 | 3 |
| Electric | 2200 | 3 | 4037 | 5 | 2455 | 4 | 2759 | 4 | 2603 | 5 |
| Plumbing | 4950 | 7 | 4529 | 6 | 3032 | 5 | 4739 | 6 | 2223 | 4 |
| Appliances | 0 | 0 | 89 | 0 | 0 | 0 | 0 | 0 | 0 | 0 |
| Heat | 2200 | 3 | 458 | 1 | 0 | 0 | 0 | 0 | 0 | 0 |
| Misc. | 1320 | 2 | 5416 | 7 | 990 | 2 | 3316 | 4 | 2186 | 4 |
| totals | 70266 | 100 | 77695 | 100 | 63215 | 100 | 76405 | 100 | 57229 | 100 |

### Square Feet of Architectural Area

The "architectural area" method described in the American Institute of Architects' book "Architectural Graphics Standards" uses the sum of a building's floor areas that are not less than six feet high, measured from the exterior wall faces and including partitions and stairwells. Areas with headroom less than six feet, porches, and basements are discounted by half.

Square foot costs are useful for general planning, but they are not precise, partly because different kinds of spaces have different costs. Bathrooms cost more than bedrooms; and if the number and quality of fixtures is the same, a larger bath will have a lower square-foot cost than a smaller one. Areas are also measured variously; if living-space floor area is priced, where are the additional costs for a space two stories high, or a basement, or garage?

### Choosing an Approach to Energy-Efficiency

In the past ten years, Community Builders has designed and built a wide variety of solar and energy-efficient homes which perform well in our 7500 degree-day climate. These have included several low-cost direct gain and indirect gain houses. We have also had extensive experience with buffered houses and superinsulation. Out of this experience, we make these recommendations for new housing in our cold climate:

• *In any house, use as much insulation as practical considerations allow, along with measures to control infiltration heat loss.* Superinsulation, which carries these conservation measures to a high degree, can cut fuel needs to a tiny amount and eliminate the need to install a large heating system. The increase in mortgage payments which results from the additional construction cost is frequently balanced by the decrease in heating costs (see discussion on the next pages). In most cases we have found that it is better to put money into conservation than into the large amounts of added glass and mass of the direct gain and indirect gain solar heating approaches.

• *Plan the solar orientation and window areas to use the sun's heat in conjunction with the thermal mass that is already present in the structure and contents of the building.* We call this approach sun-tempering. If a concrete floor is planned — which is less expensive than a wood floor — the south windows can be sized larger to utilize that additional mass.

The difference between sun-tempering and direct gain is a difference in emphasis and degree. With direct gain, the use of a large south glass area requires additional thermal mass. With sun-tempering, no extra thermal mass is used and the area of south windows is limited by the amount of heat the normal mass of the building can absorb. Sun-tempering is a basic, simple, sensible way to cut heating costs in most buildings.

• *For owners who want a greenhouse or very large glass areas for visual enjoyment, a sun/earth buffered design is unequalled.* It can protect the greenhouse from freezing without requiring added heat or storage; it can provide good interior daylighting without glare; and it requires very little auxiliary heat. All this can be accomplished for not much more than the cost of building the greenhouse. (A budget cost allowance for the greenhouse might be at the same square-foot figure as the rest of the house.)

## CONSERVATION COSTS AND SAVINGS

Construction costs that improve thermal performance may actually reduce an owner's overall expenditure for housing. Consider some typical construction costs associated with reducing the heat loss from the various parts of a 1500sf two-story house, built to four different standards of conservation.

The upper part of the chart on the opposite page shows the costs associated with increasing the level of insulation from the 1970-type house with R-11 walls to the superinsulated house with R-39 walls. Also included on the chart is the cost of an electric heating system, which decreases as the house is better insulated.

The lower part of the chart shows the corresponding heat loss rates from various parts of the house, in Btu/hr/deg F. These heat loss rates and proportions are independent of climate.

The estimated yearly auxiliary heating loads for a 7500 degree-day climate are shown at the bottom of the heat loss chart, in millions of Btus, along with the costs for electric heat at $.07/kWhr. The auxiliary heating loads are affected to a great extent by factors such as occupants' habits and weather. The estimates of heating loads were based on hourly simulations using weather data for northern New England.

Notice that the heat loss rates (in Btu/hr/deg F) are NOT directly proportional to the yearly heating costs. For example, the heat loss rate of the super-insulated house is 73 percent lower than that of the poorly insulated 1970 house, but the yearly heating cost is 90 percent lower in the given climate, and in fact the superinsulated house would require no auxiliary heat at all in many regions.

In the chart of construction costs, all four houses are assumed to be well-built, above average in quality. All costs were estimated using 1983 prices and include a contractor's mark-up. Window costs are all for a similar style and high quality, from one manufacturer. (The style, quality, and number of windows have large effects on the costs of construction.)

For clarity and simplicity, we have assumed electric resistance heat in all these calculations. There are many other choices for auxiliary heat; they are discussed later in this chapter.

In the superinsulated house, costs include an air-to-air heat exchanger large enough to provide fresh air at the rate of 1/4 AC/hr with the exchanger operating half the time.

Other assumptions:
  25ft x 30ft 2-story
  100sf south windows
  20sf windows on each of E,N,W
  air-to-air heat exchanger in the superinsulated house is 60cfm at 70 percent efficiency, operating half the time (effective .14 AC/hr)
house volume is 12000cf (1500sf x 8ft)
thermostat set at 68F day, 60F for eight hours
  at night

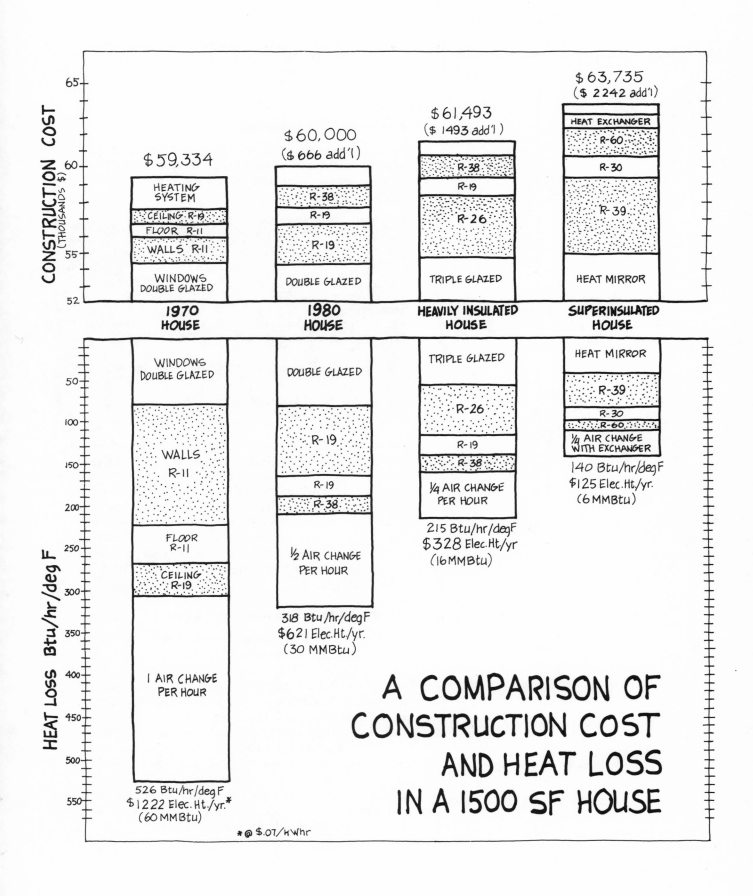

A COMPARISON OF
CONSTRUCTION COST
AND HEAT LOSS
IN A 1500 SF HOUSE

## Analysis of Costs/Savings

If we assume electric heat at $.07/kWhr, the yearly heating costs and savings for the four houses are shown below, with the additional construction costs:

|  | yearly fuel bill at $.07/kWhr | yearly fuel savings | added constr. cost |
|---|---|---|---|
| 1970 house | $1222 | $0 | $0 |
| 1980 house | $621 | $601 | $666 |
| heavily-insulated house | $328 | $894 | $2159 |
| superinsulated house | $125 | $1097 | $4401 |

This gives us the following simple payback periods:

| | |
|---|---|
| 1980 house | 1.1 years |
| heavily-insulated house | 2.4 years |
| superinsulated house | 4.0 years |

However, if we compare the superinsulated house to the heavily-insulated house instead of to the 1970 house, we have a cost increase of $2242 and a yearly fuel savings of $203, for simple payback of 11 years (instead of the 4 years above). This is one illustration of the fact that economic analysis of this type is very dependent on the given reference point. It also illustrates that the first steps toward conservation can have the most dramatic and beneficial effects.

## Combined Fuel and Mortgage Costs

Simple payback is a questionable criterion for evaluating an investment. Another way to look at the cost/savings relationship is in terms of the total cost of both fuel and mortgage, although again in the next chart we are considering only the first year costs and have made no allowance for fuel inflation. Since mortgage costs are usually paid monthly we have divided the first-year fuel costs by 12; the chart compares the average monthly combined costs of mortgage and fuel. Mortgage payments are based on a down payment of 20 percent of the total price, an interest rate of 12 percent, and a mortgage life of 30 years.

| First-year average of monthly fuel plus mortgage | | | |
|---|---|---|---|
| | monthly mortgage at 12% | average monthly fuel cost | total monthly cost |
| 1970 house | $488 | $102 | $590 |
| 1980 house | $494 | $52 | $546 |
| heavily-insulated house | $506 | $27 | $533 |
| superinsulated house | $524 | $10 | $535 |

The total monthly cost goes down for each improvement until we get to the superinsulated house, when it goes up $2 per month (actually $1.53 per month, rounded upward). That slight increase in total monthly cost would quickly be negated by either a small increase in fuel cost or a small decrease in mortgage interest rates. For mortgage rates below 11 percent, the super-insulated house shows the lowest total monthly cost even in the first year.

At a fairly modest fuel inflation rate of 5 percent per year, the average fuel cost over 30 years is more than double the initial fuel cost. Therefore the table above could more realistically appear as follows, assuming an annual fuel inflation rate of 5 percent:

| 30-year average of monthly fuel plus mortgage | | | |
|---|---|---|---|
| | monthly mortgage at 12% | 30-year average monthly fuel cost | 30-year average monthly total cost |
| 1970 house | $488 | $236 | $724 |
| 1980 house | $494 | $120 | $614 |
| heavily-insulated house | $506 | $63 | $569 |
| superinsulated house | $524 | $24 | $548 |

With fuel inflation considered, the average monthly cost over the 30-year life of the mortgage goes steadily down with each improvement in conservation; the superinsulated house has the lowest average monthly cost.

## Life-Cycle Cost

Probably the best way to evaluate an investment of this kind is in terms of the life-cycle cost of both construction and fuel, over the life of the investment. To do this, we first need to find the value, in present day dollars, of all future fuel costs. This "present value" of future energy costs is calculated using a discount rate which takes into account both interest and inflation. The 30-year life-cycle costs for the four houses are shown below, again assuming 12 percent interest and fuel inflation of 5 percent.

### Life-cycle costs, in present-value dollars

|  | construction cost | present value of 30-year fuel costs | total life-cycle cost |
|---|---|---|---|
| 1970 house | $59334 | $15691 | $75025 |
| 1980 house | $60000 | $7977 | $67977 |
| heavily-insulated house | $61493 | $4213 | $65706 |
| superinsulated house | $63735 | $1606 | $65341 |

The superinsulated house shows the lowest life-cycle cost in the table above. Keeping the rest of the assumptions constant, the superinsulated house has the lowest life-cycle cost if the fuel inflation rate is above 3.5 percent. At inflation rates below 3.5 percent the heavily-insulated house has the lowest life-cycle cost. In general, higher fuel inflation rates increase the benefits from conservation, and higher interest rates decrease the benefits.

An additional factor which has not been included in any of the figures above is the effect of deducting the interest on a mortgage from taxable income (if deductions are itemized). In the first years of a mortgage, most of the payments go just toward interest, and this savings on income tax can reduce significantly the total cost of a larger mortgage, making the life-cycle costs of conservation even lower.

## Some Conclusions

In reviewing the figures above, which are all based on a climate of about 7500 degree-days per year, these observations could be made:

• In order to do any analysis of this kind, it is necessary to make a large number of assumptions, many of which have a big effect on the bottom line figures.

• Each added layer of insulation saves less fuel than the previous layer. The first steps in conservation are usually the most cost-effective.

• The decreases, or savings, in life-cycle costs due to each new level of conservation are:

| | |
|---|---|
| from 1970 house to 1980 house: | $7,048 |
| 1980 house to heavy insulation: | $2,271 |
| heavy insulation to superinsulation: | $365 |

• The life-cycle cost of the superinsulated house is slightly lower than that of the heavily-insulated house, but the difference is small. There are other benefits to superinsulation that do not show up in the life-cycle cost figures. Some of them are mentioned below.

## Heavy Insulation vs. Superinsulation

When the total life-cycle cost of superinsulation is only slightly lower than it is with heavy insulation, why incur a greater initial expense for the relatively small savings?

• Superinsulation provides security. It provides insurance against any unanticipated increase in energy costs. For example, electric costs can make sudden large jumps (sometimes 40% or more) when expensive new power plants go on-line. Super-insulation also provides security in the event that fuel supplies are cut off entirely: a superinsulated house could be kept moderately warm with a tiny amount of heat from almost any source.

• It is very easy to maintain uniform, comfortable temperatures throughout a superinsulated house. Heat stratification is normally very small.

• There will be more days of the year when the house does not require any auxiliary heat at all — more days when there is no need to start a fire, no need to turn thermostats up and down, no need even to think about managing auxiliary heat.

## SITE

The site of a house can have a large effect on its design, and the individual character of each location is worth careful study. Many factors must be integrated if the house is to work well for its occupants and have the appearance of belonging where it is. Sometimes what initially looks like a site problem can be turned into an asset.

Site considerations include:
  Aesthetics
  Neighborhood, noise, views
  Orientation possibilities
  Solar access
  Drainage
  Driveway location and expense
  Septic system location and expense
  Water supply
  Wind protection
  Summer shade
  South-facing slope for sun warmth

It helps to make a rough map showing the important characteristics of the site, and the various ways the land will be used. Uses may include:

house, greenhouse, garage, woodshed, driveway, parking

septic system, water supply, patio, lawn, flowers, clothes yard, play area, volleyball or other sports area

vegetable garden, fruit trees, berry bushes

pond or swimming pool

trees for shade, wind protection, beauty, firewood, timber

barn, kennel, or chickenhouse, with needed runs, pasture, fences

Careful site utilization brings satisfaction and enjoyment through the years to both occupants and passersby.

Planning advice and information about agricultural characteristics of the land may be obtainable from county agricultural agents. Neighbors also may know a lot about the land.

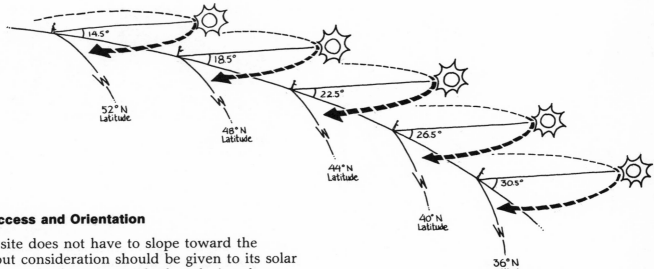

**December Sun Angles**

## Solar Access and Orientation

A good site does not have to slope toward the south, but consideration should be given to its solar access — to sunshine on south glass during the heat-producing midday hours in winter. Can the house be located so that during those midday hours trees or buildings or other obstructions will not shade much of the south glass? (Remember also that superinsulation and buffering can work without sunshine.)

To find solar south accurately in any location: at local solar noon, which is halfway between sunrise and sunset as reported in many newspapers, mark the shadow of an exactly upright pole or a string supporting a hanging weight. (Solar noon varies east to west across a time zone.) A compass indicates magnetic orientation, which in most locations differs substantially from true orientation. Even when the difference between magnetic and true direction is known, the possibility for confusion is great enough to present a real risk of ending up with a house that is facing well off true solar south.

The chart at the right shows the approximate reduction in solar gain due to turning the glazing away from south, for northern latitudes (about lat. 36 to lat. 50).

The sun's warmth is most effective at midday or solar noon, when the energy is most directly reaching the earth. At other hours, the proportions of the maximum hourly amount of solar energy available on vertical south glass in December are:

| window orientation | degrees from true south | sunshine percentage (w/vertical glazing) |
|---|---|---|
| S | 0 | 100 |
| SSE, SSW | 22 | 91 |
| SE, SW | 45 | 70 |
| ESE, WSW | 67 | 45 |
| E, W | 90 | 25 |

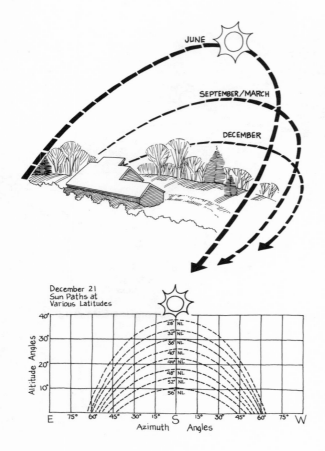

December 21
Sun Paths at
Various Latitudes

| Percent of Noontime Radiation in December | | | | | |
|---|---|---|---|---|---|
| latitude | solar noon | 1pm & 11am | 2pm & 10am | 3pm & 9am | 4pm & 8am |
| 36 deg N | 100% | 96% | 85% | 65% | 30% |
| 40 deg N | 100% | 96% | 83% | 60% | 20% |
| 44 deg N | 100% | 95% | 80% | 54% | 6% |
| 48 deg N | 100% | 94% | 77% | 43% | 0% |

To approximately determine the sunshine's access to a building site, face true south and visualize the December sun coming over the horizon in the southeast and setting in the southwest. The arc of its path between sunrise and sunset will reach a high point at solar noon that varies with the location's latitude, from 38 degrees above the horizon at lat. 28 to 10 degrees at lat. 56. Sweep your eyes across this arc, and notice any trees or buildings between the sun's path and you. Particularly note the portion of the arc that represents your peak solar hours, when the sun is delivering the most heat. This is your window to the sun.

Note which obstacles you can control; evergreen trees in this area would probably be the first to go. Even the bare winter crown of deciduous trees can block up to 50% of the available sunshine. Sometimes lower branches of close tall trees can be thinned while upper branches are left for summer shading. (Try to leave summer shade.)

For more accuracy, the altitude and azimuth (angle-from-south) of skyline obstructions can be measured and charted on diagrams of the sun's path at the site's latitude.

## DEVELOPING FLOOR PLANS

Floor plans can evolve best if the functions of the house and the relationships between functions are determined in the earliest stages of planning, before the rooms are specifically defined. What should happen where? How will people move around inside the house? How will they approach the house, and what is in the area around the house that they will want to see or use?

When functions have been considered and defined, spaces and rooms — and structure — can follow.

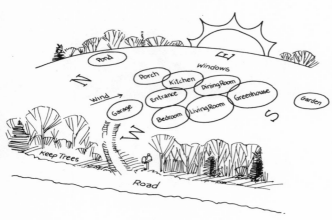

Structural systems are more economical if floor plans and dimensions are designed to use standard lumber lengths most efficiently. Framing lumber normally comes in even-foot lengths, and at the normal joist spacing of 16″ on center, 2x6's can span 8 feet, 2x8's can span 12 feet, and 2x10's can span 14 feet. So when the main structural sections (or "bays") in a house are 8, 12 or 14 feet wide, the framing lumber is used near the limits of its carrying strength, avoiding the necessity of using the next larger size of joist. Although 2x6's are not commonly used as floor joists, their use in 8ft bays can be very practical and economical.

Simpler rectangular building shapes cost less than arrangements which have more corners and larger perimeters. The outside surface area of a more cube-shaped building is smaller than one with a more elongated or complex shape, reducing both the cost of construction and the surface area heat loss of the occupied house. These cost considerations must be balanced with style preferences and aesthetic choices.

**Evolution of Plans**

We start with roughly sketched designs — floor plans and sections — which take account of budget costs and size, and incorporate all that is identified of the activity and space relationships. 1/8″ grid graph paper under tracing paper permits quick and accurate locating of building elements without measuring.

Once the basic requirements of a design have been established, it helps to allow time for the plans to grow. Model-building can make the plans seem more real. As plans grow, continue to think about design, structure, cost, thermal functions and gravity airflows, traffic, furniture space, storage, light and general atmosphere.

Try to define everything, because everything affects something else. Sometimes a design element which seems to work in rough outline can result in unacceptable complications when the details are filled in. Innovations may have hidden interrelated effects.

**Root Cellars**

A useful space that seldom receives aware planning is a root cellar for winter storage of fruits and vegetables. It should be thermally separated from the rest of a basement and have its own ventilation. An earth floor is usual, or concrete without a vapor barrier.

## INSULATION

Effective insulation and infiltration control are crucial elements in energy-efficient construction (see later for air and vapor barriers). Any house can employ many of the same techniques for reducing heat loss that are used in superinsulated houses: thicker walls with multiple insulation layers, continuous vapor barriers, triple-glazed windows.

The amount of insulation used depends partly on the anticipated temperature differences across the insulation. A wall between livingroom and greenhouse would require less insulation than a direct wall between livingroom and outdoors, and still less if greenhouse temperatures are to be kept high. A direct wall in a warm climate of course requires less insulation than a direct wall in a cold climate. The optimal amount of insulation in part of a house also depends on the house's total approach to conservation: superinsulated walls would not be cost-effective in a house with many double-glazed windows or with little ceiling insulation. Refer back to the chart earlier in this chapter of four houses insulated to various degrees, and notice their various balances of insulation.

### Types of Insulation

Fiberglass is the most commonly used insulation material. It is easy to work with and provides the highest R-value per dollar. It must be kept dry with effective vapor barriers since even small amounts of moisture reduce its insulative value significantly. Friction-fit material is preferred, not faced with paper, so that installers can easily see that the insulation fills the spaces completely. Friction-fit material is also more dense and more resistant to heat flow. Blown-in fiberglass can be effective if a uniformly firm density is achieved.

Cellulose (made from shredded newspapers with fire-retardant and other chemicals added) can be blown into cavities or over ceilings. It also must be protected from moisture with a vapor barrier. Its insulative quality is also dependent on how densely and uniformly it has been installed.

Many types of rigid foam insulation are now available. Most are relatively impermeable to moisture and some can function as vapor barriers.

To some extent they are fire or smoke hazards and should be protected with gypsum if used inside a house. When used outdoors they must be protected against surface damage including that caused by ultraviolet rays from the sun.

The higher R-values of the more expensive foams do not keep pace with their higher cost, and it is often more economical to use a greater thickness of a cheaper insulation to achieve the same total R-value.

Rigid insulation panels — sandwiches of gypsum or flakeboard skins over foam insulations — make effective unit walls and roofs that can work well outside exposed timber frames. Their chief advantages are ease of construction (keeping labor costs down) and the continuous tight quality of the resulting insulating shell. Their chief disadvantage is their relatively high material cost.

We generally limit our use of rigid foam insulation to foundation exteriors, night window insulation, and vent doors because it does not usually seem cost-effective elsewhere.

## Night Window Insulation

Night insulation at exterior glass areas can easily cut these windows' heat losses in half. The heat loss through a square foot of even a triple-glazed window is about 8 times the loss through a square foot of an 8-inch insulated wall. In buffered houses, most glass areas have small heat losses because they are exposed to the greenhouse or other buffering airspaces, so there is not a great need there for night window insulation, but it can be cost-effective at window areas directly between the living space and the outdoors. It is also useful on greenhouse exterior windows if higher nighttime temperatures are wanted in the greenhouse.

The simplest and least costly movable insulation may be snugly fitted polystyrene or urethane/isocyanurate sheets pressed into window openings; disadvantages are the storage space required and potentially hazardous fumes in case of fire. Edges can be protected with 2″ duct or foil tape, and faces can be painted or covered with paper or cloth. Some foil-faced boards are available with a decoratively-embossed surface.

### Rigid Insulation at Foundation

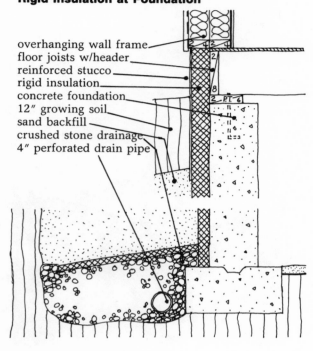

overhanging wall frame
floor joists w/header
reinforced stucco
rigid insulation
concrete foundation
12" growing soil
sand backfill
crushed stone drainage
4" perforated drain pipe

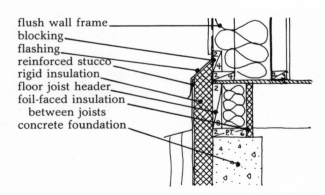

flush wall frame
blocking
flashing
reinforced stucco
rigid insulation
floor joist header
foil-faced insulation
   between joists
concrete foundation

### Drainage

Positive perimeter drainage, with crushed stone or other durable protection against silting, is important for dry basement floors. Backfill above the footing drain is sand or gravel, up to the level of grass or plant roots. Although it is usually one of the last steps in construction, the first and most important part of good drainage is being sure that the grading of the ground surface is sloped to cause rainwater to run away from the perimeter of the house. Good earth surface drainage away from building walls will also reduce the load on foundation drains.

Many other systems are described in the books "Thermal Shutters and Shades," and "Movable Insulation" listed in the reference section. Space provisions for night insulation may affect window or wall design.

### Attic Access and Basement Doors

Uninsulated attic spaces often have very little insulation at a stairway door or ceiling access panel. If the attic space is to be used for storage, insulated covers or doors can be designed. If not, an exterior gable access door can eliminate the need for a direct ceiling connection and reduce heat loss between house and attic.

Basement bulkhead doors give little thermal protection, and usually need to be supplemented with a well-gasketed interior insulated door.

### Foundation Insulation

Insulation is used on foundation walls to reduce heat loss to the outdoors from a basement, crawlspace, or slab floor. Foundation insulation is most important above grade and near the earth surface, since winter temperature differences are larger here than in the deeper earth, and therefore heat losses are higher. The amount of insulation may decrease as depth increases.

Expanded polystyrene (EPS, or "beadboard") has had a long record of durable effectiveness as foundation insulation, even in wet clay soils. Extruded polystyrene is also recommended for this use, but when durability and long-term cost efficiency are both considered EPS provides a higher R-value per dollar. (Information is available from EPS suppliers or the Society of the Plastics Industry, 3150 Des Plaines Avenue, Des Plaines, Illinois 60018.)

An additional horizontal skirt of insulation extending outward (and somewhat downward, to drain) from the bottom of the vertical wall insulation will further increase the distance that heat must travel to get from crawlspace or basement to outdoors. If this outer rigid insulation skirt is not evenly supported on well-tamped fill it is likely to be broken by the weight of the overfill.

Insulation above grade should be protected from abrasion and from ultraviolet degradation by stucco, aluminum flashing, cement-asbestos board or other opaque materials. We usually use a cement stucco coating reinforced with a polyester mesh.

If termites could be a hazard, a metal termite shield should protect the house framing, and the stucco should extend to the footings.

Our wall framing usually overhangs the foundation several inches to provide space for insulation under it, but when the wall does not extend beyond the foundation a cap flashing can cover the insulation top.

If a house foundation is on ledge which has little earth cover, and the ledge under the house is to be used as a source of low-temperature heat, earth-protected insulation can extend outward from the foundation on the surface of the ledge.

Placing insulation at the outside of foundations rather than on the inside allows the mass of the foundation to be used as part of the thermal storage of the basement or crawlspace. If the wall mass is not considered important for heat storage, the insulation can be on the inside — although vapor barrier continuity at floor joists and headers may then be much more difficult to achieve.

Inside the foundation, rigid insulation can be nailed or cemented to the inside of the concrete, and gypsum board cemented to the insulation. Alternatively, wood furring can be nailed to the concrete, or a separate interior wood frame wall can be built and insulated with fiberglass.

## WOOD FRAME WALL SYSTEMS

The buffered house and the superinsulated house can use various framing systems, several of which are described here. Some additional aspects are discussed in the Construction chapter.

Basic to all are a tight WIND BARRIER near the outside surface, and a VAPOR BARRIER close enough to the inside surface to keep moisture vapor from condensing within the wall and harming insulation, framing, or finishes. These barriers are discussed later in this chapter.

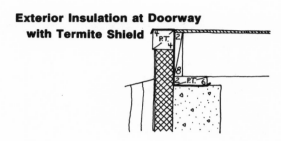

**Exterior Insulation at Doorway with Termite Shield**

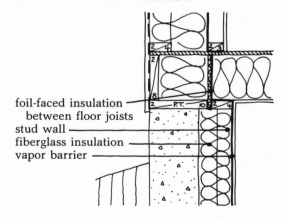

**Interior Foundation Insulation**

foil-faced insulation between floor joists
stud wall
fiberglass insulation
vapor barrier

**1— Post and beam w/infill and girts**

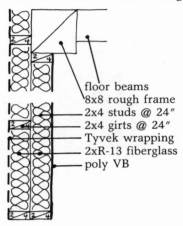

floor beams
8x8 rough frame
2x4 studs @ 24″
2x4 girts @ 24″
Tyvek wrapping
2xR-13 fiberglass
poly VB

**2— Post and beam w/4-1/2″ panels**

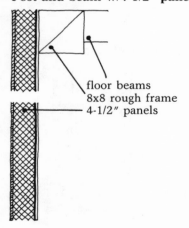

floor beams
8x8 rough frame
4-1/2″ panels

Illustrations here show the wind barrier as a heavy dashed line and the vapor barrier as a heavy solid line.

1983 costs (including labor) and nominal resistances to heat loss are shown for each wall system, for an average square foot of exterior wall framing. The costs include insulation and wind or vapor barriers, but no finishes, inside or out.

These cost figures are intended for use in a general comparison and evaluation of the various systems, but they would not be realistic everywhere, since costs of both material and labor vary in different places and under different conditions.

Walls as defined here — just the wall framing and insulation, with infiltration and vapor barriers — might normally account for five to seven percent of total house costs.

POST AND BEAM is the oldest of these framing systems. Beams are large enough to carry floor and roof loads between posts. Walls are non-structural, filling between posts or hung outside. (We are assuming 24-inch spacing of light framing in this and other examples.) The character and atmosphere of exposed framing is usually the reason for this choice, unless the builder has very low-cost access to suitable timbers. A well-fitted smooth-surface oak frame could cost much more than the rough cut softwood we have assumed here.

#### #1 — Post and beam with 2x4 infill and 2x4 girts

Typical cost:        $3.25
R-value:             26

#### #2 — Post and beam with sandwich panel

Typical cost:        $4.23 (excluding value of gypsum)
R-value:             25

SINGLE STUD wall systems are less costly than post and beam because they require less material, using the same pieces, the studs, to accomplish spacing, stiffness, and load-bearing.

### #3 — 2x8 Stud wall          (2x6:)

Typical cost:      $2.05     ($1.69)
R-value:           26        (19)

Rigid polystyrene or foil-faced urethane or isocyanurate board insulation can be added on inner stud faces. The foil facing can provide an effective vapor barrier if edges are caulked and joints sealed with foil tape. Most rigid insulations are too impermeable to moisture vapor to be used safely on the outside of framing.

Typical cost of 1″:     $.47-.61
R-value:                5-7

### #4 — 2X6 Stud wall with rigid insulation

Typical cost:      $2.31
R-value:           26

WIRE-CHASE walls add a 2 or 3 inch space inside the vapor barrier, not only to provide a space for electrical work where it will not puncture the vapor barrier, but also to reduce the danger of other holes being made in the vapor barrier, particularly during gypsum installation. This wire chase can be added to almost any wall system. If a reflective-surface (foil) vapor barrier is used, the air space can add about 2 or 3 to the R-value of the wall; without a reflective surface the airspace adds about R-1. The wire-chase space can also be filled with insulation (but see discussion of the possibility of moisture condensation in the next section).

### #5 — Wire chase inside 2x6 wall with rigid insulation

Typical cost:      $2.61
R-value:           29

CROSS-HATCH framing is especially efficient when vertical exterior siding will be used. This system combines vertical load-bearing studs on the inside (usually 2x4's) with horizontal girts or nailers on the outside (frequently 2x4's, with 2x8 window and door frames, or 2x6 with 2x10 frames). Diagonal bracing is nailed to the inner framing members.

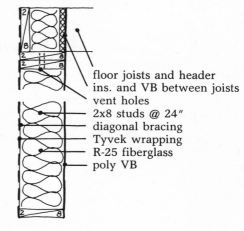

3— 2x8 stud wall

floor joists and header
ins. and VB between joists
vent holes
2x8 studs @ 24″
diagonal bracing
Tyvek wrapping
R-25 fiberglass
poly VB

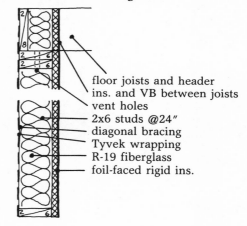

4— 2x6 stud wall w/rigid ins.

floor joists and header
ins. and VB between joists
vent holes
2x6 studs @24″
diagonal bracing
Tyvek wrapping
R-19 fiberglass
foil-faced rigid ins.

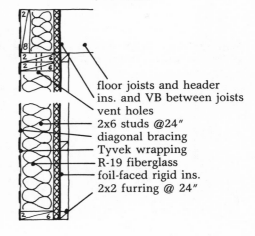

5— 2x6 wall w/wire chase

floor joists and header
ins. and VB between joists
vent holes
2x6 studs @24″
diagonal bracing
Tyvek wrapping
R-19 fiberglass
foil-faced rigid ins.
2x2 furring @ 24″

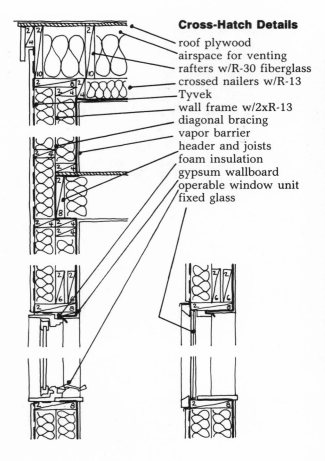

### Cross-Hatch Details

roof plywood
airspace for venting
rafters w/R-30 fiberglass
crossed nailers w/R-13
Tyvek
wall frame w/2xR-13
diagonal bracing
vapor barrier
header and joists
foam insulation
gypsum wallboard
operable window unit
fixed glass

### Plan of Cross-Hatch Wall at Corner and at Bearing Wall

Flat 2x4s over roof purlins provide a ventilation space under roof plywood.

## Outrigger Details

- roof plywood
- vent space
- insulation stops
- cellulose insulation
- plywood gussets
    - hang outrigger
    - below truss
- Tyvek
- 2x4
- fiberglass insulation
- vapor barrier
- plywood sheathing
- 2x4 wall
- plywood gusset
- 2x4 block
- PT plywood soffit

## Plan of Outriggers on Standard 2x4 Walls

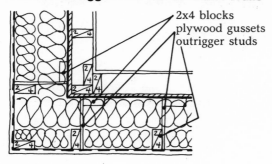

- 2x4 blocks
- plywood gussets
- outrigger studs

## Plan View of Larsen Truss Outside Standard Frame Wall

## Double-Stud Details

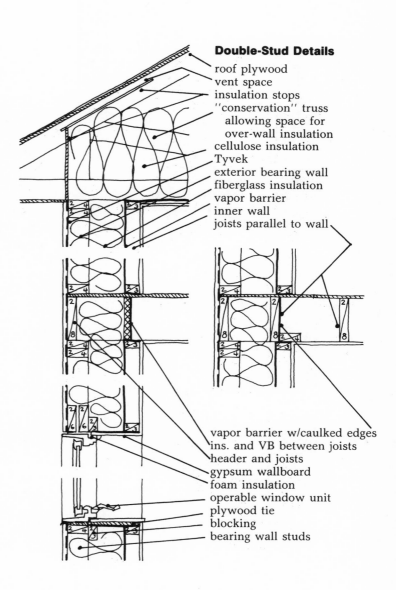

- roof plywood
- vent space
- insulation stops
- "conservation" truss
    - allowing space for
    - over-wall insulation
- cellulose insulation
- Tyvek
- exterior bearing wall
- fiberglass insulation
- vapor barrier
- inner wall
- joists parallel to wall
- vapor barrier w/caulked edges
- ins. and VB between joists
- header and joists
- gypsum wallboard
- foam insulation
- operable window unit
- plywood tie
- blocking
- bearing wall studs

## Plan of Double-Stud Wall at Corner and at Bearing Wall

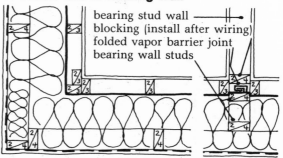

- bearing stud wall
- blocking (install after wiring)
- folded vapor barrier joint
- bearing wall studs

**6— 8″ cross-hatch wall**

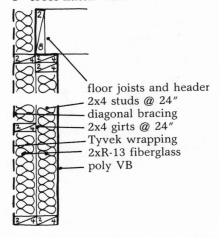

floor joists and header
2x4 studs @ 24″
diagonal bracing
2x4 girts @ 24″
Tyvek wrapping
2xR-13 fiberglass
poly VB

**7— 8″ double-stud wall**

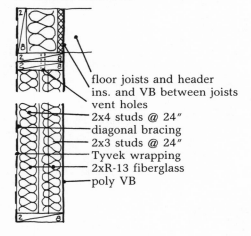

floor joists and header
ins. and VB between joists
vent holes
2x4 studs @ 24″
diagonal bracing
2x3 studs @ 24″
Tyvek wrapping
2xR-13 fiberglass
poly VB

**8— 12″ double-stud "Quickwall"**

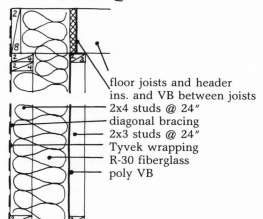

floor joists and header
ins. and VB between joists
2x4 studs @ 24″
diagonal bracing
2x3 studs @ 24″
Tyvek wrapping
R-30 fiberglass
poly VB

### #6 — 8″ Crosshatch wall    (10″:)

| | | |
|---|---|---|
| Typical cost: | $2.00 | ($2.36) |
| R-value: | 26 | (32) |

DOUBLE STUD systems have separate inner and outer vertical framing members. This slightly reduces heat transfer directly through the framing, since insulation is about six times less conductive than wood. (Using the same amount of insulation, an 8-inch wall framed with staggered 2x4's will have an average R-value about 2 higher than the same wall framed with 2x8's.) Another advantage of a double-stud system is that the inner and outer framing can be separated as much as desired to provide more space for insulation without adding any more studs. Labor costs are higher than for systems with fewer pieces, but the savings in materials makes this a very cost-effective way of achieving a high R-value.

In thicker walls, vapor barriers can be placed on the outside face of inner wall studs, thus providing a wire chase.

In the "Saskatchewan" system, inner and outer walls are built and tipped up together. When the outer studs are load-bearing, the inner walls are usually put up later, after windows are in and the roof is tight, as in the "Quickwall."

Figures in parentheses below are for walls with insulation added to the wire chase space (middle figure) or by increasing the wall thickness (figure on right).

### #7 — 8″ Double-stud wall

| | |
|---|---|
| Typical cost: | $2.11 |
| R-value: | 26 |

### #8 — 12″-15″ Double-stud "Quickwall"

| | | | |
|---|---|---|---|
| Typical cost: | $2.32 | ($2.61) | ($2.55) |
| R-value: | 31 | (42) | (39) |

### #9 — 12″-15″ "Saskatchewan" unit wall

| | | | |
|---|---|---|---|
| Typical cost: | $2.75 | ($3.04) | ($2.98) |
| R-value: | 31 | (44) | (39) |

The OUTRIGGER starts with a standard plywood-sheathed frame wall and can be used for retrofit superinsulation of an existing house. Vapor barrier sheeting is applied to the wall exterior, and preassembled outrigger frames are nailed to the walls. The frames as used by Gerald Leclair in Winnipeg, Saskatchewan, consist of vertical 2x4s with plywood gussets and nailer blocks for connecting at floors and roof. (A "Larsen Truss" by John Larsen of Edmonton, Alberta, uses two 2x2s with gussets every two feet. Addresses for purchasing additional information on both systems are in Reference Section D on page 211.) Cost here includes the standard sheathed wall.

### #10 — 8" Outrigger on 2x4 wall

| | | | |
|---|---|---|---|
| Typical cost | $2.84 | ($3.13) | ($3.07) |
| R-value: | 31 | (44) | (39) |

## HUMIDITY AND VAPOR BARRIERS

Loose houses with a high rate of infiltration tend to be very dry in the winter, since there is a constant influx of cold outside air and when this cold air is heated, its relative humidity becomes very low. Well-insulated tight houses do not have this large source of cold dry air, so the moisture created by the natural occupancy of the house tends to build up, creating higher levels of relative humidity.

This can be a more healthful atmosphere for people, but in cold climates the exterior building shell stands between warm moist air and cold dry air, and the shell must be designed and constructed with great care.

Warm humid air in a house cools as it moves out of the structure toward the cold outside, and at some point in that cooling process it will reach the dew point (100 percent relative humidity) and vapor in it will condense. In an insulated wall, this dew point will typically be reached somewhere within the insulation. The resultant build-up of water can cause serious problems, particularly the deterioration of insulation and the destruction of paint and wood.

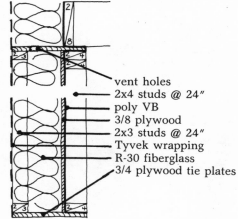

**9— 12" "Saskatchewan" unit wall**

- vent holes
- 2x4 studs @ 24"
- poly VB
- 3/8 plywood
- 2x3 studs @ 24"
- Tyvek wrapping
- R-30 fiberglass
- 3/4 plywood tie plates

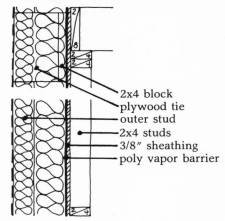

**10— Outrigger on standard wall**

- 2x4 block
- plywood tie
- outer stud
- 2x4 studs
- 3/8" sheathing
- poly vapor barrier

### Relative Humidity and Dew Point

The amount of water vapor that a volume of air can hold is dependent on the air temperature. The RELATIVE HUMIDITY (RH) of air is the percentage of water contained in the air compared to the maximum amount that the air could hold at its particular temperature. Warm air can hold much more water vapor than cold air can hold. The relative humidity of a volume of air containing a particular amount of water decreases as the air is heated, since the maximum amount that the air can hold is increased. Conversely, the relative humidity increases as air is cooled.

The DEW POINT is the temperature at which air is cold enough to have a relative humidity of 100 percent. When further cooled, some of the water vapor condenses to form liquid water.

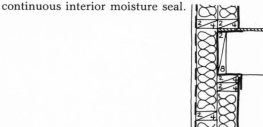

A strip of vapor barrier hangs over an exterior wall while floor joists are nailed to the header. One side of it will then be wrapped around the outside of the header and the other later joined to a lower wall vapor barrier to maintain a continuous interior moisture seal.

This attic truss space will be ready for blown-in insulation after a vapor barrier and gypsum board ceiling have been installed. At the eaves, the plywood baffles will contain insulation and fiber spacers will assure venting. Insulation has been installed in the outer stud spaces of the wall below but not between the inner studs.

## Wall Protection

There are two basic approaches to preventing condensation build-up in walls. The first is to prevent or slow the movement of moist air into and through the wall, with air and vapor barriers (which may or may not be combined in one material). The second is to allow or encourage the outward migration of whatever water vapor is in the wall, so moisture can not build up. This kind of ventilation is now fairly standard practice at roofs, but it has not been commonly used at walls.

Somewhere near the inside surface of the wall, exterior wall systems should provide a continuous vapor barrier, as free of penetrations as possible. Maintaining good vapor barrier integrity requires careful planning in the design stage as well as careful installation. An example is shown in the illustrated detail which allows a continuous vapor barrier where second-floor framing meets outside walls.

Interior walls, even at a sunspace or greenhouse, normally need no vapor barrier because temperatures within the wall would not get as low as the dew point and condensation would not occur.

## Exterior Sheathing

Vapor barriers near the interior surface of exterior walls help keep air moisture in the living space and out of the walls. However, it is very difficult to maintain perfect vapor barrier continuity, and the inevitable flaws in the vapor barrier may still cause condensation problems if low-permeance sheathing materials prevent the escape of moisture vapor to the outdoors. Plywood and chipboard sheathing have such high resistance to the relief of moisture vapor pressure that they are difficult to use without a large risk of moisture damage within the wall. Most rigid insulations, faced or unfaced, are also relatively impermeable. Whenever impermeable materials — foils, rigid insulation, plywood, or chipboard — are used on the outside of exterior walls, venting channels should be provided, either to the attic or more directly outdoors.

Many building suppliers now stock Tyvek, an infiltration barrier fabric which blocks the flow of air while it allows moisture vapor to pass through. Wrapping the outside of a wall with nine-foot wide sheets is an easy and effective way to keep wind out of wall insulation. Since Tyvek is a flexible material it does not provide any bracing; bracing is usually accomplished with steel straps. With the use of Tyvek on the outside and a good vapor barrier on the inside, air movement through the wall is kept to a minimum.

## Location of Vapor Barrier in Wall

Since a vapor barrier's function is to stop the passage of air-borne moisture so that it cannot condense in the wall, it is crucial that the vapor barrier be located inside the point in the wall where condensation would occur. The exact point in a wall where condensation could occur is dependent on the temperature and humidity inside and outside, and on the R-values and permeance of the various materials in the wall.

A general rule of thumb for moderately cold climates is to have at least three times the R-value on the cold side of the vapor barrier as on the warm side. There is no danger of this type of condensation if the vapor barrier is immediately behind the interior finish material, but if substantial insulation is to be used on the warm side of the vapor barrier, the R-value ratio should be checked.

A few objective investigations of walls that have been in service for several years have seemed to indicate that improperly located vapor barriers do not always cause problems, but there have been so many real and undeniable instances of moisture damage caused by improperly located or insufficient barriers that good judgment certainly calls for caution and care.

The ratio of the outer R-value (on the cold side of the vapor barrier) to the inner R-value (on the warm side of the vapor barrier) is computed from the R-values of the different layers, and is shown below for several wall systems. The layers are shown from the outside in, and the location of the vapor barrier is indicated by "VB".

Tyvek infiltration barrier wraps the house walls.

**fg:** fiberglass batts (or combination of batts)
**VB:** vapor barrier (polyethylene for top part of chart,
inner foil surface of urethane for bottom part)
**EPS:** R-4 expanded polystyrene "beadboard," unfaced
**urethane:** R-6 board, with foil surface used for vapor
barrier
**airspace:** app. 2″ space, R-2.5 with reflective surface,
R-1 without

| wall system (outside → inside) | Total R-value | R-value ratio (outer to inner) |
|---|---|---|
| 4″ fg, VB, 4″ fg | 28 | 1.0 to 1 |
| 6″ fg, VB | 21 | 17.7 to 1 |
| 6″ fg, VB, airspace | 22 | 9.4 to 1 |
| 6″ fg, VB, 1″ EPS | 25 | 3.9 to 1 |
| 6″ fg, VB, 4″ fg | 34 | 1.4 to 1 |
| 8″ fg, VB | 28 | 23.9 to 1 |
| 8″ fg, VB, airspace | 29 | 12.7 to 1 |
| 8″ fg, VB, 1″ EPS | 32 | 5.3 to 1 |
| 8″ fg, VB, 4″ fg | 41 | 2.0 to 1 |
| 10″ fg, VB, airspace | 35 | 15.5 to 1 |
| 10″ fg, VB, 1″ EPS | 39 | 5.4 to 1 |
| 10″ fg, VB, 4″ fg | 47 | 2.3 to 1 |
| 12″ fg, VB, airspace | 42 | 18.8 to 1 |
| 12″ fg, VB, 1″ EPS | 45 | 7.8 to 1 |
| 12″ fg, VB, 4″ fg | 53 | 2.8 to 1 |

In the following systems, the foil face of the urethane
(with foil tape) creates the vapor barrier:

| | | |
|---|---|---|
| 4″ fg, 1″ urethane, VB | 21 | 17.7 to 1 |
| 4″ fg, 1″ urethane, VB, airspace | 24 | 5.5 to 1 |
| 4″ fg, 1″ urethane, VB, 4″ fg | 34 | 1.4 to 1 |
| 6″ fg, 1″ urethane, VB | 27 | 23.0 to 1 |
| 6″ fg, 1″ urethane, VB, airspace | 30 | 7.2 to 1 |
| 6″ fg, 1″ urethane, VB, 4″ fg | 40 | 1.8 to 1 |
| 8″ fg, 1″ urethane, VB, airspace | 37 | 9.1 to 1 |
| 8″ fg, 1″ urethane, VB, 4″ fg | 47 | 2.3 to 1 |
| 10″ fg, 1″ urethane, VB, airspace | 43 | 10.7 to 1 |
| 10″ fg, 1″ urethane, VB, 4″ fg | 53 | 2.8 to 1 |
| 12″ fg, 1″ urethane, VB, airspace | 50 | 12.7 to 1 |

All systems in the chart assume these additional outer
R-values:

| | |
|---|---|
| exterior air film | .17 |
| wood siding | .80 |

and these additional inner R-values:

| | |
|---|---|
| 1/2″ gypsum | .45 |
| interior air film | .68 |

The degree of protection obtained with various R-value ratios is shown in the next chart. At 70F inside, the R-value ratios in the left-hand column provide protection against condensation on the inside of the vapor barrier down to the temperatures shown in the two right-hand columns, for 50 percent and 70 percent RH (relative humidity) inside.

| R-value ratios outer to inner | at 50 percent indoor RH, protected to these outdoor temperatures | at 70 percent indoor RH, protected to these outdoor temperatures |
|---|---|---|
| 1 to 1 | 28F | 50F |
| 2 to 1 | 7F | 40F |
| 3 to 1 | -14F | 30F |
| 4 to 1 | -35F | 20F |
| 5 to 1 | -56F | 10F |
| 6 to 1 | -77F | 0F |
| 7 to 1 | -98F | -10F |
| 8 to 1 | -119F | -20F |
| 9 to 1 | -140F | -30F |
| 10 to 1 | -161F | -40F |

If condensation occurs in wall insulation only for brief periods during the coldest weather, the water can presumably evaporate and disperse safely when the wall is warmer again.

## WINDOWS
### South Glass

Most of the factors affecting the choices of area and location of south glazing are considered in later separate chapters on buffered and superinsulated houses.

Because glass is such a poor insulator, the amount used has a large effect on a building's heat losses and gains. In general, increasing the area of south glazing results in increased temperature swings. A space inside a larger glazing area will tend to get both hotter on sunny days and colder at night than it would with less glazing.

### East, North, and West Glass

Glazing amounts and location on the east, north, and west are determined more by considerations of

High glazing close to the sloping ceiling brings light far back into the livingroom of the Burns House.

daylighting, view, summer ventilation, appearance, and safety than by the concern for the best heating performance. Windows on these sides lose much more heat in winter than they bring in, so for optimal winter thermal performance a house would have no such windows at all. This extreme is almost never acceptable, but the combined area of windows other than on the south is often kept fairly small.

If west windows are not shaded they can bring in too much summer sun in the afternoon, when it is least welcome.

In cold climates, at least three glazing layers are recommended for windows on the east, north, and west. Quadruple-glazing or night insulation can further reduce heat loss, and also reduce condensation on the glass.

## Daylighting

In most rooms it is possible to provide daylight from at least two directions. Since glass areas can account for perhaps a third of the heat loss from a house, it is worthwhile to try to make the best use of every square foot.

— Windows close to a light-colored wall can bathe the wall with light and add much brightness to a room.

— Splayed jambs (the slanting sides of a window opening, in which the inner opening is larger than the outer) in a thick wall soften the contrast between indoor and outdoor light levels and make a window seem larger.

— An enlarged light-colored sill — a light shelf — can project the light from a window deeper into the room.

One of the advantages of a buffered design is the possibility of using the buffer space for bringing daylight to inner rooms without glare, overheating, or large nighttime losses. These possibilities sometimes are unrecognized and some buffered houses have had windows only on the south, even when they had a double north wall. Balanced lighting from different directions is much more pleasant.

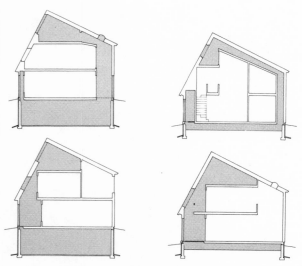

Sections of Rich and Winnisquam houses show how light from the buffer space or a skylight reaches into these houses.

Skylights have been used since ancient times to enhance daylighting. A light shaft can be used to bring light from a skylight down to a lower story. Exterior reflectors can even be used to increase the amount of light transmitted by a skylight without increasing the area of the skylight (which would increase the heat loss). Heat loss through skylights is also reduced by using multiple glazings.

### Glazing Details

Glazing must keep out rain, keep in heated air, and control both vapor moisture and condensed moisture.

Moisture condensation between layers of multiple glazing is usually prevented by a desiccant sealed by the manufacturer between the sheets of glass. Tiny (1/16″) relief slits to the exterior can also keep the space between two individual sheets of glass largely free of condensation, although less conveniently, less certainly, and possibly more expensively. If sheets of glass are individually installed, it is the inner glass that is most tightly sealed because moisture-vapor movement is from the higher-humidity interior toward the drier exterior. Occasional removal of the unsealed outer glass may be necessary for cleaning.

Glass is used much more than any other glazing material. It is durable and economical and stays clear indefinitely, unlike many other glazing materials.

Acrylic is obtainable (on special order) without an ultra-violet absorber, to obtain a broader sunlight spectrum in a greenhouse. Acrylic glazings undergo large expansion and contraction when they change temperature, and require specially designed gasket materials which permit the glazing to move.

A glazing fundamental refined in many years of manufactured greenhouse designs is the "shingling" of water-carrying surfaces, so water running down off one surface will follow a planned safe course from surface to surface down to where it can acceptably collect or drip.

Lighting in this photo comes entirely through the buffer spaces; note that the diningroom window at right rear is closed off with a Window Quilt. Winnisquam House.

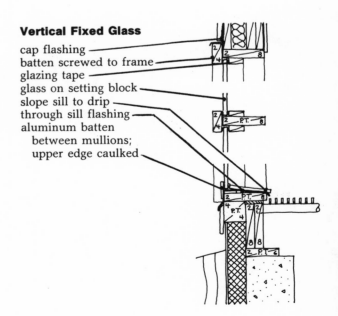

**Vertical Fixed Glass**

cap flashing
batten screwed to frame
glazing tape
glass on setting block
slope sill to drip
through sill flashing
aluminum batten
  between mullions;
  upper edge caulked

## Fixed Glass in Roof

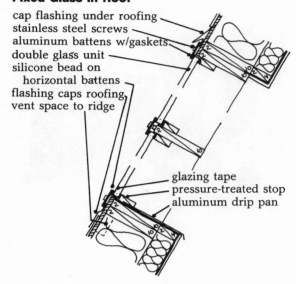

cap flashing under roofing
stainless steel screws
aluminum battens w/gaskets
double glass unit
silicone bead on
  horizontal battens
flashing caps roofing
vent space to ridge

glazing tape
pressure-treated stop
aluminum drip pan

## Condensation on Windows

In tight, well-insulated houses with high humidity levels, multiple glazing layers are necessary to prevent condensation on exterior windows which could drip down and damage wood and plaster. The level of protection against condensation which is provided by different numbers of glazing layers is shown in the chart below, for an indoor temperature of 70F. The column for 70 percent RH illustrates the difficulty of preventing condensation when the indoor humidity is very high.

| number of glazing layers | at 50 percent indoor RH, clear above these outdoor temperatures | at 70 percent indoor RH, clear above these outdoor temperatures |
|---|---|---|
| single-glazing | 43F | 57F |
| double-glazing | 13F | 43F |
| triple-glazing | -16F | 29F |
| quadruple-glazing | -45F | 15F |

## FRESH AIR

Designers and builders of all types of energy-efficient houses have become increasingly aware of the necessity for good infiltration control in keeping heat losses low. By caulking all cracks, using tight-sealing windows and doors and carefully constructing continuous vapor and wind barriers, it is possible to build a house with a very low rate of infiltration and air change. This is very desirable in terms of heat loss, but it also means that there may not be enough natural ventilation to provide an adequate amount of fresh air for the occupants.

Fresh air means oxygen for respiration, it means air relatively free of harmful pollutants, and it means air free of unpleasant odors.

Some Pollutants:

Carbon monoxide and nitrogen oxide result from the use of unvented gas appliances and from tobacco smoking.

Formaldehyde is a component of the bonding agents used in plywood and particle board and also in carpeting and upholstery, and escapes from them as a gas.

Radon is a radioactive gas naturally present in some soils and rock (and consequently some concrete). It enters the air from such materials under, around, and inside the house. It has been estimated that indoor exposure to radon may be responsible for 1,000 to 20,000 deaths from lung cancer each year in the U.S.

Particulates — in smoke, dust, and lint — can also be harmful.

### Danger of Negative Pressurization in Tight Houses

Deadly monoxide gases from a woodstove or other fuel-burning appliance may be pulled into the livingspace if a house becomes depressurized enough to create a reverse flow in a chimney flue. Depressurization in a tight house can be caused by exhaust fans, dryers, or by air-to-air heat exchangers operating in frost-control mode (when the air intake is stopped or slowed). Reverse chimney flow can also sometimes occur with an uninsulated exterior chimney or under certain wind conditions.

Operable windows or air-to-air heat exchangers can be used in place of some exhaust fans. Open earth tubes, discussed later in this section, could provide dependable access to outside replacement air.

If exhaust fans or vented dryers are used when a fuel-burning appliance is operating, a window should probably be opened to supply replacement air.

Smoke detectors of the ionization type or gas detectors can provide backup safety.

### Air-to-Air Heat Exchangers

To provide fresh air for occupants, windows to outdoors can always be opened for ventilation, but leaving a window open for fresh air tends to defeat the original purpose of tightening the house.

To keep air quality high and the heat loss low in a tight house, fresh air can be brought in through an air-to-air heat exchanger. In this a pair of fans push outgoing and incoming air past each other in adjacent but separate channels, so that the heat in

"Air changes per hour" (AC/hr) has been used to define the quality of air in buildings, but clear and simple standards for an acceptable fresh air input rate are not easily established. Some reports have indicated that even in houses with air change rates below 1/4 of an air change each hour the indoor air quality remained satisfactory, but most recommendations have been for from 1/4 to 1/2 or more of an air change per hour. The actual rate at which fresh air is needed to replace polluted air depends on the nature and amount of pollutants, either generated in the house or entering from outside. Occupants' habits clearly have a very large effect on indoor air quality. Odors are annoying but frequently harmless.

the outgoing air can preheat the incoming air. These devices typically recover 50-70 percent or more of the heat which would otherwise be lost when warm inside air is replaced with cold outside air.

Heat exchangers can be cost-effective, since the energy used to power the fans is is much less than the energy which would be required to heat the same amount of air entering a house directly from outside in cold weather. The savings that result from keeping the rate of natural infiltration in a house very low may result in a quick payback of the initial cost of an air-to-air heat exchanger.

Occupants' lifestyles affect not only the production of pollutants but the choices about where and when the pollutants are removed, as from kitchen, bath, and laundry, and from rooms where people smoke. The effectiveness of a heat exchanger in removing pollutants from any particular house is also very much affected by the amount and location of the mixing of fresh and room air, whether with one ductless unit in a wall or ceiling or a unit having ducts to or from several parts of the house.

Some air-to-air heat exchangers recover moisture and the latent heat that moisture represents in addition to sensible heat. Others do not recover moisture, and although they may be slightly less efficient they have the advantage of functioning additionally as de-humidifiers by replacing the moist inside air with drier outside air. This de-humidifying effect is often very desirable in a tight house with high humidity.

Selecting and locating a heat exchanger wisely requires consideration of many factors, which are well discussed and indexed by William A Shurcliff in "Air-to-Air Heat Exchangers for Houses."

### Heat Loss Due to Infiltration

The chart below shows how big a part infiltration can play in the total heat loss from a house. Three houses are shown, with air change rates from 1 AC/hr down to 1/4 AC/hr. The left-hand column under each house shows the percent of heat loss from natural infiltration, and the right-hand columns show the percent of heat loss if the house were completely tight and the same amount of fresh air was brought in through a 70% efficient air-to-air heat exchanger.

| **Percent of house heat loss due to infiltration** | | | | | | |
|---|---|---|---|---|---|---|
| | 1980 insulated house (1) | | heavily-insulated house (2) | | super-insulated house (3) | |
| AC/hr | natural | w/exch. | natural | w/exch. | natural | w/exch. |
| 1.00 | 51% | 24% | 57% | 29% | 66% | 37% |
| .75 | 44% | 19% | 50% | 23% | 60% | 31% |
| .50 | 34% | 13% | 40% | 17% | 50% | 23% |
| .25 | 20% | 7% | 25% | 9% | 33% | 13% |

(1) 1980 insulated house: R-19 walls and floor, R-38 ceiling, double-glazed windows

(2) Heavily-insulated house: R-26 walls, R-19 floor, R-38 ceiling, triple-glazed windows

(3) Superinsulated house: R-39 walls, R-30 floor, R-60 ceiling, triple-glazed heat mirror windows

All four houses:
25ft by 30ft, 2-story
100sf glass on south, 20sf on each of E, N, W
Volume: 12000 cubic ft

Without an air-to-air heat exchanger, a natural air change rate of .5 AC/hr represents forty percent of the total heat loss in the heavily-insulated house. Even a rate of .25 AC/hr without an exchanger represents 1/4 of the losses from the heavily-insulated house and 1/3 of the losses from the superinsulated house; with the same amount of air brought in through an exchanger the infiltration losses are reduced to 9 and 13 percent, respectively, of the total house losses. In the super-insulated house an air change rate of 1 AC/hr, which is typical for much recent conventional construction, would amount to 2/3 of the house heat loss.

## Earth Tubes

Another strategy for providing fresh air in tight houses which has been frequently proposed but little tested is to bring air from the outside through long underground tubes, so that the air is heated to the temperature of the earth before it enters the house. Such tubes might be 40-70 feet long with a diameter of 6 inches, buried 4 feet or more underground.

Norman Saunders comments: Earth tubes are optimally multiple 4″ tubes on roughly 3ft centers as deep as they can be put (but above the highest water table), and with the outside entrance up a hill near roof height. Such are good for about 100 Btu/day/foot. Unless the exterior entrance is well above the floor you may need a small fan.

Larger earth tubes have also been used (with questionable effectiveness) for summertime cooling.

Earth tubes should slope to drain. A gooseneck inlet should be higher than rodents could reach.

The key problem in the use of an earth tube to provide winter fresh air is the seasonal cooling of the earth around the tube as cold outside air is being constantly drawn through it. The earth around the tube can become so cold that it can no longer usefully heat the air.

William Shurcliff has proposed an improvement to this scheme to eliminate the seasonal cooling problem and provide greater efficiency. Two tubes with fans are installed, with the fans always running in opposite directions, one pulling air into the house while the other blows air out. The fans are set to reverse direction periodically, so that the heat in the outgoing air is used to warm the earth around the tubes. This could counteract the seasonal cooling of the earth, and allow for recovery of some heat from the exhausted air.

## AUXILIARY HEAT

To maintain comfortable temperatures all winter in a cold climate, even very efficient houses still often require some heat in addition to solar and internal gains. Conventional heating systems are much larger in output than is necessary for such houses and would not be cost-effective.

### Wood Heat

Small woodstoves currently seem to be the most frequent choice for auxiliary heat in very efficient houses. Our survey of envelope homes showed 3/4 of the owners using wood for their primary source of auxiliary heat.

There is a charm in seeing and feeling the center of warmth that a woodstove creates in a house, and many people find satisfaction in the tasks of gathering wood and tending a stove. The local availability of firewood in many areas of the country can make it a very economical and secure source of fuel.

However, if a house requires little supplementary heat, a woodstove will have to be small, and may require more frequent tending than a larger stove in a less efficient house. Some owners have eliminated the stove entirely, and the chimney, because they would have been used so little. A chimney is a sizeable investment, adding $1000 or more to the cost of a house (unless one is already needed because a fireplace is planned).

Woodstoves require more tending than most other types of auxiliary heat, and chimneys need occasional cleaning. Woodstoves can contribute significantly to indoor air pollution if improperly managed.

Banks and other funding institutions may not approve mortgage money for a house unless some heat source other than a woodstove is planned, even though a small stove may be more than adequate. The installation of electric resistance heaters is usually the least expensive way to satisfy a bank in such a situation.

Conventional fireplaces are very inefficient, and in fact can contribute a net heat loss even when fired up, because of the large volume of air and heat which goes up the chimney. The addition of tight-fitting glass doors to a fireplace cannot be expected to increase the net efficiency, because the glass blocks much of the radiant heat that the fireplace gives off. If a fireplace is expected to provide heat, it should have a heat exchanger and probably a small fan. We have not yet seen a two-way fireplace of any type that produces effective heat. One solution for people who want a fireplace is to have a fireplace for pleasure and a woodstove for heat. Another solution is to use a woodstove that can be opened like a fireplace for special occasions.

A good source of information on all aspects of wood heat is the "Solid Fuels Encyclopedia" by Jay Shelton.

### Electric Heat

Electric resistance heaters and radiant panels are frequent choices for auxiliary heat sources because of their convenience, their automatic control of temperature, their flexibility of location, and their low initial cost. They require little or no floor space, and are particularly useful in houses with many closed separate areas, where dependable warmth and sound privacy are both desired.

The operating cost of electric heat is often three or four times that of fossil fuel or wood heat. However, the amount of auxiliary heat needed in a very efficient house is so small that the cost may still seem low. An indirect cost associated with the use of electric heat for houses comes from the need for utility companies to build expensive additional power plants to maintain peak power capacity for the coldest weather periods.

Frequently, when some other type of heat is the main choice in a house, electricity is still used for certain areas such as bathrooms, or for backup at times when the use of other fuels would be inconvenient, such as when owners are ill or away.

### Gas and Kerosene

Small space-heat units for gas and kerosene are available for installation on exterior walls where combustion gases can be inexpensively and safely vented through a metal or plastic pipe. Fuel costs are usually much lower than for electricity, and the wood-burner's costs of a masonry chimney are avoided. These heaters are becoming increasingly popular solutions to the need for auxiliary heat in very efficient houses. Another possibility, where air circulation throughout the house is prevented by many normally-closed doors, is a conventional gas domestic hot water heater (with a flue) coupled to baseboard or other radiators, with a circulating pump.

### Stratification

One of the many advantages of very well-insulated buildings is their tendency to maintain uniform temperatures throughout the house. Stratification of air in poorly-insulated houses is made worse by the large differences in air temperature between areas where much heat is being lost and the areas (such as around a high-output stove) where a lot of heat is being supplied to replenish those losses. In a well-insulated house the losses are much smaller, and the amount of heat supplied to replace losses is small, so the air maintains much more uniform temperatures. Warm air rises only because there is cooler air around it to create a difference in temperature. A high-temperature heat source such as a woodstove will tend to create more stratification than a source of heat which is more spread around, but when heat loss rates are very low the extent of stratification is still small.

In very well-insulated houses, natural air movement will often provide enough distribution of the heat from a single heat source, but very large houses and houses with more than two heated stories usually require more than one source of heat.

Heat distribution can be enhanced by providing airflow openings between rooms, especially if openings are located where temperatures and air density differences will be greatest. Floor grilles under second-floor windows can permit the naturally-falling cooler air at those locations to push warmer air up a stairway, for example. The value of the heat distribution that such airflow openings provide will need to be balanced against the loss of sound privacy.

**Combustion Air**

Combustion air is required by any woodstove, fireplace or other oxygen-consuming appliance, and should be considered in planning ventilation. A small to medium woodstove requires about 10 to 50 cfm (cubic feet per minute) of outside air, which is equivalent to about .05 to .25 AC/hr (air changes per hour) for a 1500sf house. An open fireplace could require 20 times as much, or up to the equivalent of 5 AC/hr.

These figures suggest that a fireplace in a tight house should have a duct to bring outside air in near the fireplace opening. For a woodstove, the existing ventilation in a house may provide air for combustion as well as fresh air for occupants; if not, a duct can be used or a window can be opened slightly as necessary. We have used 2″ ducts for woodstoves in tight houses. The use of a combustion air duct for a stove does not necessarily increase the overall efficiency, but it can be convenient and reduce drafts. Any combustion air duct should include a way to close it when not in use.

## FIRE SAFETY

Fire safety is a critical consideration in the design of any house. Provisions for prevention and control are both necessary. Houses requiring very little auxiliary heat may be safer from fire than houses with harder-working heating systems, but they still contain most of the same hazards.

Fire requires three elements: combustible materials, oxygen, and an ignition source. The absence of any one can prevent or stop a fire.

Housekeeping with an awareness of potential fire hazards makes a difference, keeping combustibles away from possible ignition sources.

Early detection of excessive heat or smoke is critical if people are to escape; almost all fire deaths are from heated, toxic, or oxygen-deficient products of combustion, which can build up quickly. Heat and smoke detectors should be located at each bedroom ceiling and at all spots where they can best pick up heat and smoke information: at high points — including tops of stairways and airflow spaces — and near places where fire might start. Detectors should be interconnected so all will sound when any one is activated. Battery backup should be provided, with failsafe automatic recharging. (Fans and gas or oil heating equipment should automatically shut off.)

Quick and dependable means of escape for occupants is important; each sleeping room requires an alternate escape route for use if fumes or heat are in the central hall. Sleeping rooms are safest when each is a totally separate compartment with no nighttime air connection to other rooms.

Containment of fire spread is aided by surfacing rooms and air spaces with gypsum board or plaster which delay fire penetration and spread.

Spring-loaded dampers can be installed to block airflow spaces when high temperatures melt a restraining fusible link. When they work properly (sometimes they have not) they can effectively block fire travel, but they are more practical for small ducts than for whole-wall airspaces. We have not used them or considered them appropriate in the situations we have faced. If dampers are to be used, it is important that they be selected at the

### Fire in Envelope House

The importance of fire safety awareness was highlighted by a 1983 fire in Aurora, Colorado which destroyed an envelope house which had a double north wall open to the tops of bedroom ceilings.

The fire started in an attached garage next to the double north wall of the house. Although the fire disabled the electric service on which fire detection equipment depended, occupants noticed smoke, left the house, and called the fire department. Firemen put out the fire in the garage, and thought there was no fire in the house, not knowing of the double wall in which the fire was still smouldering.

When firemen opened an upstairs bedroom window, intending to clear the house of smoke, the north wall space became a chimney and within a minute the entire house was engulfed in smoke and flame. The fire spread so rapidly that a fireman on the second floor was temporarily overcome even though wearing breathing apparatus. The house was quickly destroyed.

time that airflow spaces are being designed, and their use should include some means of ensuring their continuing dependability after installation.

When a dependable water supply is available — not from a single well where electric failure caused by a fire would interrupt the pumping of water — simple temperature-activated sprinklers may provide control.

## SUMMER COOLING

The way a house is managed can make a large difference in comfort during hot weather. Much outdoor summer heat can be kept from entering a well-insulated and well-shaded house, simply by closing windows when unwanted warmer air could come in. If nights are cool enough, windows and doors can be opened then to cool the house in preparation for the next day's heat.

Summer earth temperatures are enough lower than maximum inner house air temperatures to permit effective use of the earth as a heat sink. A fan-forced circulation of inner house air down to a basement or crawlspace, or through pipes buried in the earth, can remove significant amounts of heat from the living spaces. Underground cooling tubes that bring outdoor fresh air into the livingspace are much less effective, even if they have a large surface area.

Shading windows from summer sun will help keep down cooling loads.

# 5

## BUFFERED HOUSE DESIGN

Many of the concepts and details of buffered house design are the same as for other energy-efficient houses, and have been discussed in Chapter 4. In this chapter we will cover the elements more specific to the buffered house, including the greenhouse/sunspace, airflows, and the earth connection.

There is no best way to design a buffered house. It is clear now that houses can please their owners and have very small needs for auxiliary heat with little south glass or with a lot of south glass, with gravity airflows or with fans, in climates with little sun or in climates with much sun, and in many shapes and sizes. More than anything else, the survey of "envelope" houses (Chapter 9) shows that many different strategies and configurations in envelope house design have produced remarkably good results.

Some of these houses also have problems, or have had problems that required correction, and while it seems inappropriate to set limits on the total design, we can write about particular details that work, and about the principles that underlie success. Understanding the functions and requirements of the elements that go into the whole — insulation, buffering, earth connection, sunshine, convection, storage — can lead to the effective application of these many-sided concepts that can be combined in so many different ways.

A house is also more than a formula-built structure. From the many subtleties of its microclimates to the owner's daily choices in its management, each house is unique.

## THE GREENHOUSE/SUNSPACE

The greenhouse/sunspace is particularly individual and organic. Owners with varied interests, skills, and backgrounds experiment, observe, and learn, and their greenhouses constantly change.

There are two main choices in function of the greenhouse/sunspace in a buffered house, besides its thermal functions:

— To grow plants, either food-producing or ornamental;

— To provide an additional living space closely connected to the outdoors.

The relative importance of these two functions to the owner will have a large effect on the sunspace design. For some owners, the beauty of a solarium and its function as an additional living space are more important than the capacity for plant growth; others want a greenhouse that will provide substantial amounts of food during much of the year. Many will want a sunspace that is part greenhouse and part solarium. The first step in greenhouse design is for prospective owners to identify the relative importance of these functions to their own lives.

Visiting with other greenhouse/solarium/sunspace owners will help in picturing possibilities and give an opportunity to ask questions. Both beginners and veterans enjoy and gain much from visiting and talking with other owners; greenhouse conferences and classes are often available.

### South Glazing

One basic fact about the amount of greenhouse glass is that increasing the glazing area will make the space both hotter on sunny days and colder at night — although high and low temperatures are also greatly affected by airflows and by the available thermal storage mass. Some of the factors affecting the choice of amount and location of south glazing in a buffered house are:

• *The intended main use of the greenhouse:* With a moderate amount of vertical south glass, a greenhouse can provide light and a view and produce wintertime salad greens, but if the greenhouse is to produce a substantial amount of food year-round, a larger glazing area and probably some sloped overhead glazing are required to give plants the light they need.

• *Climate:* In locations that tend to have long wintertime cloudy periods, overglazing can make a greenhouse too cold for people and plants during those sunless times — which also increases heat loss from the inner house through windows between the interior and the greenhouse. In locations that have a high percentage of wintertime sun, designers have more freedom in the amount of glazing used. Auxiliary heat or thermal shades can compensate somewhat for a lack of dependable sunshine, but with added cost.

• *The use of a natural convective airflow:* If the glazing area is very large or high up, it is unlikely that natural convection will achieve the airflow rates that are necessary to move heat quickly from the greenhouse into storage. If the airflow is insufficient, the greenhouse may get too hot on sunny days.

• *The use of added thermal storage materials:* In houses that use added thermal mass to warm livingspace floors or provide other higher-temperature uses of greenhouse heat, a fairly large south glazing area may be necessary in order to produce the required heat. It may also be necessary to use fans to move the heat to the storage location.

• *Aesthetics:* Since the beauty of a greenhouse is one of its primary attractions, aesthetic consider-ations will play an important part in decisions regarding glazing amount and location.

Envelope houses built in the past few years typically have had greenhouse glazing areas equivalent to 15-30 percent of the inner house floor area — and have experienced great differences in temperature and airflow.

The designer should try to be aware of just where the heat collected behind south glazing is going to go. If there is insufficient storage mass (whatever the forms of storage) or if transfer into storage is insufficient, overheating will result.

## Placement of Glazing

Keeping greenhouse/sunspace glazing down low

gives more push to a gravity circulating flow, since air heated low in a sunspace rises farther than air heated near the top. High glazing, because it creates more heat while adding little push, contributes to stagnation, and its use for light or vision will tend to cause larger temperature swings — unless a fan is used.

A sloping south roof provides space for overhead greenhouse glazing, and a surface for future photovoltaics (although a shallow slope could be too much covered with snow). The space under sloped glass can be an excellent location for a domestic hot water collector (see later section in this chapter). Summer heat gains are less easily controlled with sloping glass, but good ventilation can prevent overheating.

Where plant production is emphasized, glazing with as much height as the south-north depth of the greenhouse will throw light to the back wall for additional planting beds or trays. In solariums, some shading toward the back may be more pleasant.

Where living area views are through the green-house to the outdoors, vertical glass will be cleaner and have fewer reflections than sloped glass. As with all windows, horizontal crossbars should not be at eye level for people sitting or standing.

## Reducing Heat Loss Through Glazing

Double-glazing is almost always used for the greenhouse. One glazing layer would increase the heat loss through the glass too much; three or more layers would significantly reduce the light available to plants (and cost more than the heat it would save).

Most owners of buffered house do not use movable insulation, preferring the convenience and simplicity of leaving their buffering glass uncovered. As a consequence, minimum temperatures in the greenhouse — and in the crawlspace or basement — are lower than when night insulation is used. Many plants can survive or do well in the lower temperatures; people wait for the sun's warmth before they use the room.

Window Quilts in North Carolina home.

Photo: New Energy Window Systems

## Growing Food

Plant growth requires light for photosynthesis. For this reason, a food-producing greenhouse needs some overhead, sloped glazing, except perhaps in far-north latitudes. With vertical glazing, plants can be grown in the area near the glass where light is brightest, but some sloped glazing lets in much more light in spring, fall, and summer. A light-reflecting white surface on the north wall will also increase light on plants, and help keep them from turning too much toward the south. Light from the east will give plants an earlier morning warm-up and more hours of daylight. Light from the west is less important and the west side is often fully insulated.

Some plant varieties need more light than is available in winter with either vertical or sloped glazing, and would require artificial light, from lamps providing the spectrum plants require. Standard daylight-color fluorescent lamps are suitable. Adequate light helps keep plants healthier as well as growing faster; plants with inadequate light are more susceptible to invasion by aphids.

During cold weather periods, the greenhouse can get too cold for many varieties of plants; the large glass areas which are needed for good growing light also cause greater temperature swings. The greenhouse can be kept warmer by using night window insulation, by storing more sun heat to be used at night, or by supplying supplementary heat during the colder times. The most frequently-used method of providing some heat for the greenhouse during very cold periods is to leave a door or window open to the inner house, which of course cools the inner house or increases its use of auxiliary heat.

Plant varieties differ in their need for light, moisture, and heat. Owners who want a food-producing greenhouse should read about greenhouse horticulture in more than one book and talk with other greenhouse owners. Eventually they must decide whether to provide all the conditions needed by harder-to-grow varieties or to accept more easily-achieved conditions and grow plants that can withstand colder nighttime temperatures or less light.

**Earth Bed on Crushed Stone Floor**

(all wood pressure-treated)
steel tie strap to 2x4 plate
2x4 posts @ 4ft
plank sides and bottom
growing soil
crushed stone
perforated drain pipe
2x6 on flat 2x4

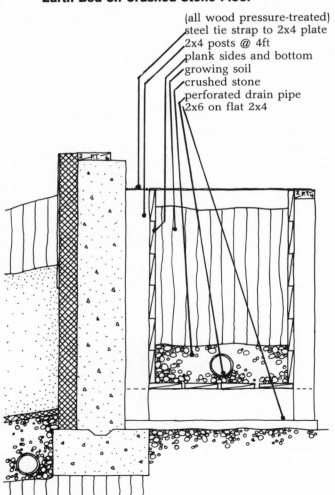

Much of the art of managing a greenhouse is in observing and adapting to the changing specific needs of plants in the varied microclimates within the greenhouse. (The reference section contains a list of food plants that have been productive in New Hampshire greenhouses and some suggestions about greenhouse management.)

### Greenhouses in Very Cold Climates

In climates colder than about 8000 degree-days or with little winter sunshine, the large glass areas of a greenhouse present greater challenges.

Extending the growing season is easy, but keeping plants growing through the winter will require not only additional measures to reduce heat losses through the glass, but also supplementary heat and artificial light during the coldest and darkest weather.

### Planting Beds and Greenhouse Floors

Growing beds with a soil depth of 24" or more will provide more natural growing conditions and require less frequent watering than shallower beds. Good drainage is essential.

In the out-of-doors, plant roots can range widely and deeply in search of the nutrients they need. In the greenhouse, plants are in a very circumscribed area, yet we are seeking maximum productivity in that small area. Soil in the beds should be the highest quality available.

Where food production is of primary importance, growing beds will be located and sized for maximum use of light, with wide beds near the window, a narrow aisle for access, and beds or space for flats on shelves to the rear with perhaps a small work area for potting.

Providing a few inches of airflow space between the earth beds and glazing will permit nighttime cold air from the glass surfaces to fall past the plants directly to the greenhouse floor and basement or crawlspace. This helps prevent freezing and provides air circulation for healthy plant growth.

Warmth in the soil is more important to plants than warmth in the air. One way to keep plant roots warmer is to use a small fan in a 4" plastic pipe to take warm air from the upper part of the greenhouse and blow it down under the growing bed soil, through a 4″ perforated drain pipe surrounded by crushed stone. Another way is to water plants with sun-warmed water.

Greenhouse floors can be earth; concrete, brick, tile, stone or wood on the earth; or any kind of suspended construction, either solid or partially open. Air from the crawlspace or basement under the house can come into the greenhouse from its north side or by passing under the greenhouse floor to come up near the south glass.

## Vertical Location of Greenhouse

The vertical location of a greenhouse can vary from cellar to attic with some advantages or challenges to each:

• Cellar: Two advantages of a greenhouse at cellar level (or between cellar and first story) are the encouragement of natural air circulation by the lower glazing location, and the space-saving convenience of having the cellar area handy for storing materials and doing messy potting jobs. Little structural support is needed for earth beds. This location is most likely to be feasible on a south slope.

• First floor: This is the most often chosen location, with the greenhouse glass buffering large livingroom windows. Greenhouse plants can be visually enjoyed from principal windows. The greenhouse can do double duty as an entryway to the house. The greenhouse is an easy few steps away for picking herbs or greens for meals or for stepping into at odd moments to tend the plants.

• Second story or attic: A rare choice, yet it can have special delights and advantages including higher temperatures to meet the needs of warmer-climate plants. A normal roof, glazed, provides the needed light. An upper story greenhouse has the privacy that makes it a special retreat. It does have special needs: forced air circulation, night shutters, care to prevent water damage below, and extra structural support for earth beds. For an example, see the Rich House described in Chapter 8.

South wall and sunspace deck from above and below.

The "Nature House" in Sweden, designed by Bengt Warne, with a glass and acrylic envelope surrounding much of the house, has greenhouses at three levels. Bengt Warne reports that the first-story greenhouse is a cold-climate growing area and is used for hardy greens; the second-story greenhouse has a milder climate and is used for moderately hardy plants; the attic greenhouse has the equivalent of a Mediterranean climate and can grow such plants as tomatoes, cucumbers, melons, and figs — by using extra heat and light to bring the plants through the dark and cold of the Sweden winter.

### Greenhouse Interior Glass

The amount of glass used between the greenhouse and the interior is based primarily on aesthetics. Double-glazing rather than single-glazing is usually required to keep down inner house heat losses. At least some of the windows between greenhouse and inner house should be operable for opening on sunny days.

The inner south glass is usually shaded enough by its location in the building to prevent the overheating of interior spaces by direct gain, but if the sunspace is very narrow and the inner south glass is so close to the outer that it would transmit too much direct gain to the inner space, heat-absorbing (tinted) glass can be used there, to help keep the heat in the sunspace. Some direct gain in the inner house may be enjoyed for the brightness and and morning warmth it gives.

### Condensation on Glass

Humidity levels are likely to be high in greenhouses and connected buffering airspaces, and in cold weather condensation will normally occur on windows between these spaces and the out-doors. When temperatures are sufficiently low, the moisture will freeze. Condensation is a natural and unavoidable consequence of the wintertime cooling of warm moist air.

Greenhouse air at 80F on a sunny afternoon might have a relative humidity of 40%, but when it cools at night it could easily reach 100% RH and the moisture would condense.

Condensation does not usually occur on glass between the greenhouse and the inner house because there is not enough cooling there.

Greenhouse humidity can be kept lower and condensation on glass reduced by using a dehumidifier or an air-to-air heat exchanger that does not recover moisture. Plants will have fewer problems with fungus and mold if the daytime relative humidity is usually kept below about 60%.

**Preventing Moisture Damage**

Moisture provides conditions for wood and paint decay, and if wood is to stand in moisture it must be specially treated chemically. Surface treatment with preservatives is not sufficient; for deep penetration wood must be impregnated while under pressure. Some of the most commonly used treatments (creosote, pentachlorophenol) are harmful to plants and people, but copper naphthanate and some water-borne preservatives are relatively safe. Surface treatment can be used on wood that is less constantly wet; the end grain should be especially well soaked in the preservative.

Condensate tends to collect on greenhouse window sills. Sills can have a drip edge and be sloped to drain away from the glass, or metal pans or gutters can be used to collect or lead water away from where it could damage plaster or wood.

Thermal-break aluminum operable windows and doors in exterior walls of greenhouses or other high-moisture spaces require much less maintenance than wood units, since they do not absorb moisture. Gypsum wall surfaces should be held back 3/16″ or more from the metal, because even with their thermal breaks the metal frames are more conductive than wood, and moisture may condense on and in the gypsum (and stain or soften the gypsum if its edge touches the aluminum). The space can be filled with glaziers' backer rod and appropriate caulk.

Glass can be coated with a chemical film which causes sheeting and eliminates the formation of water droplets, but this does not prevent condensation.

**Aluminum Window at Gypsum Wallboard**

aluminum frame w/thermal break
foam caulking
polyethylene backer rod
acrylic sealant
gypsum w/metal edge

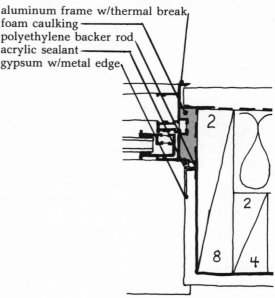

## Greenhouse Venting

Venting of greenhouse heat is usually required in the summer, and often in the spring and fall, especially if there is much unshaded sloping glass. Venting of excess heat should not be necessary in the winter, if airflows are adequate and if there is enough thermal mass.

For good plant growth, greenhouse vents, high and low combined, normally should not be smaller than 1/10 the area of south glazing.

Venting is ideally provided at both east and west ends, with some up high if possible; high vents at gable ends usually function very dependably. Clerestory vents don't always work well: they can be negatively affected by heat from roof surfaces and by winds blowing against the vents.

Ventilation and air movement across plants are helped by having some low vents also. Weatherstripped metal doors, sliding glass doors, and operable windows are usually the least expensive low vents. They may need to be secure against unwanted (burglar) entering, as well as rain.

In many buffered houses, vents are opened just once a year, in the spring, and closed again each fall. In some houses there will be days when the greenhouse would overheat in the daytime if there were no ventilation but would get too cold for plants at night if the vents were open. In these cases, an alternative to thermostatically operated fans or manually operated vents is heat-pistons — connected to open hinged sash vents automatically, when higher temperatures expand the operating material inside the piston.

Louvers and screens on vents keep rain and insects from entering, but they also reduce the effective airflow area. One simple winter closure is a close-fitting piece of mattress-type urethane foam with an inner vapor barrier and fire-protection face. Some means of holding it in place against sudden wind pressures is required.

Wind turbines are vents, not fans. They look busy and can be effective as vents if properly sized, but they do not pull air.

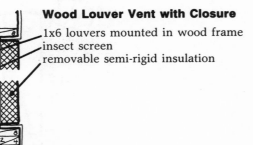

**Wood Louver Vent with Closure**

1x6 louvers mounted in wood frame
insect screen
removable semi-rigid insulation

## AIRFLOWS

Airflows in a buffered house have two main functions: to remove fairly large amounts of sun heat from the greenhouse on sunny days and, at night, to move enough heat from the crawlspace or basement to the greenhouse or other buffering airspaces to keep them from getting too cold.

In a buffered house, it is the earth beneath the house that usually provides the backup thermal mass for keeping greenhouse plants alive at night and during cloudy periods. To make use of the low-grade heat contained in the earth, it is essential to allow airflow from the basement or crawlspace into the greenhouse at night. This is facilitated by providing spaces for a convective airflow loop which makes it easy for cooled air to fall down from cold glass surfaces and for warmer earth-heated air to rise from the basement or crawlspace.

When much of the heat collected in a greenhouse during sunny hours is transferred to the interior of the house, it is not available for use in keeping the greenhouse warm when the sun is not shining. But if the greenhouse can draw on the long-term store of low-grade heat in the earth under the building (as well as under the greenhouse itself) it does not need to save much of the daily-collected heat for itself.

### Fans and Gravity Airflows

The most immediately effective means of moving a lot of air from the sunspace on sunny days is with a fan or blower. Fans cost little and can deliver more air more quickly than can gravity convection. Photovoltaic power may soon be cost-effective for powering the fan when the sun shines.

However, in our homes we usually plan for natural convection to do the air-moving. We prefer the silent and invisible forces of gravity convection, and have experienced their effectiveness. Even though air has little mass and heat-storage capacity, natural convection can move enough air to carry useful amounts of heat, if appropriate spaces and conditions are provided. Inadequate spaces for a natural airflow can cause overheating in the greenhouse, which then might require the addition of a fan.

## Airflow Spaces for Natural Convection

Because we are usually working for an effective gravity airflow, we keep airflow space considerations in mind during the early stages of design.

To transport warm air into storage, envelope houses have often used a north airflow space extending across the entire east-west length of the house. In our experience with those houses, an 8″ space along most of the length of the house seemed to be an adequate minimum dimension (adequate if surfaces could be reached for cleaning or maintenance). A presumed speed of this gravity airflow at 40 fpm (2/3 fps, or about 800 cfm) would permit a flow volume large enough to carry about half the heat produced inside 300sf of south glass. (Assumed: 1/4 heat lost back out glass, 15-20 degree drop around loop.)

A double shell or buffering airspace is most effective for saving heat at large glass areas where heat loss is high. It is less justifiable economically in solid wall areas, where less expensive systems of insulation and infiltration barriers can minimize heat loss.

When an above-ceiling airspace and a double north wall are not needed for reducing heat loss, a less costly air duct can be used, less spread out and closer to a square in shape, having fewer square feet of surface. The use of a more nearly square airflow space that has an equivalent flow area also lends itself more easily to other uses, such as for an "airlock" entry or vestibule. For example, a four foot by five foot entry provides the same flow area as an 8″ space across a thirty foot wall, with less friction for the air.

It is also possible to have effective airflow with no ducts at all, letting natural convection circulate air through the inner house itself through rooms, halls, and stairways (for example, see the Clayton house in Chapter 8).

The limits in size and shape of gravity convection airflow spaces are not easily defined. Our approach to the design of these spaces has been mostly intuitive, and we have even suggested to others that they think of themselves as being the moving air: what spaces and conditions would they want to

### Fire Dampers

Some building codes now call for fire dampers to be used "in all concealed spaces." Effective dampers at top and bottom along a whole north wall would be difficult to accomplish and probably prohibitively expensive.

find? Presumably smooth surfaces and rounded corners, and "enough" space at all parts of the loop.

Framing should be planned to keep airflow channels mostly free of framing penetrations, for less resistance to flow, for a less flammable structure, and for a more continuous vapor barrier. In airflow spaces such as an entry where people must walk, steel grating can be used to support a load without restricting the airflow. (Avoid anything unsightly in the space under the grating, as it may be highly visible.)

Fire safety should be considered when planning airflow spaces. All should be accessible. We often use natural wood surfaces in greenhouses and entries, but elsewhere avoid using exposed flammable materials, lining airflow spaces with 5/8" fire code gypsum board. (See Fire Safety discussion in Chapter 4.)

### EARTH CONNECTION

In a buffered house, the free flow of heat from deep earth to buffering airspaces during cold weather helps to ensure that buffering airspaces will not become too cold, even during long cloudy and cold periods. This earth connection is more effective if there is no insulation at or near the earth surface under the house.

Some houses have used insulation a short distance beneath the earth or basement-floor surface. This allows the near-surface mass to get warmer on sunny days but decreases the connection to a deep earth store of low-grade heat. In climates with a lot of dependable sunshine, it is possible that sufficient short-term heat can be stored in foundations and concrete floors to make a deep earth connection unnecessary. But in most climates it is the huge storage mass available through the earth connection that keeps the buffering spaces from becoming too cold under severe weather conditions when heat stored in the building materials has been used up.

Heat transfers between the earth and air are affected primarily by the amount of surface area. Conduction of heat through the soil varies with its moisture content. A moisture barrier placed near

**Steel Grating at Sliding Glass Door**

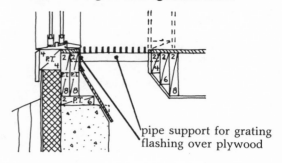

pipe support for grating
flashing over plywood

Sunspace floor of spaced 2x6 and steel bar grating.

## Mass

In many passive solar houses beautiful stone or brickwork is important in reducing temperature swings in the living space. It can do some of the same in a buffered house, but since the buffering climate is already limiting the possibility of large temperature swings in the living space this function of the stone or brick is less needed and less cost-effective.

## Backflow Dampers

A simple backflow damper of 1/2 mil plastic will prevent nighttime reverse thermosiphoning of warm house air to the greenhouse.

the surface of the soil will not only reduce humidity in the air but will keep the soil below the barrier moist so it will retain a high conductivity.

In a basement with a concrete floor, a moisture barrier should be used under the slab.

The conductivity of concrete on earth is very nearly the same as the conductivity of dense earth alone, so in terms of the earth connection it makes very little difference whether the floor of the basement or crawlspace is concrete on earth or just earth. Crushed stone under the basement slab should be avoided when not required for drainage because the airspaces between the stones would insulate the floor from the earth and reduce conductivity. A moisture barrier at the surface of the earth that created pockets of dead air would have a similar effect.

Much of the heat transfer to the earth is in the form of radiation from the ceiling of the basement or crawlspace. A reflective ceiling surface would prevent this radiative transfer and foil should be avoided.

## AIR FROM GREENHOUSE TO HOUSE

Operating strategies make some performance differences in these buildings. When greenhouse/sunspace temperatures are higher than those of the inner-house livingspace, one of the best places to store some of the sun-produced heat is in the walls, ceiling, floors, and furnishings of that living space. Opening doors and windows between the greenhouse and interior rooms during sunny hours can permit the sun-heated air to circulate into the living spaces. Not only does this provide immediate warmth to these spaces; more of the sun's heat is then stored where it is most directly available for warming people at other times, and where it is less quickly lost to the outdoors.

If no one is at home during the day to open and close connecting doors to the greenhouse, little of the sun's heat can reach and warm the interior. Temperature-activated heat pistons can be used to open hinged vent doors, or a thermostatically-controlled fan can be installed, in which case there should also be a separate return-air opening with a pressure-operated damper.

With a lot of air exchange between house and greenhouse, fresh outdoor air may be brought into the buffering airspace instead of directly into the house. This can allow some pre-heating of outside air before it enters the inner house.

Exhaust fans from the kitchen or bathroom are sometimes vented to the buffering airspace instead of directly to the outside, in order to conserve heat. A clothes dryer also can be vented to the buffer space in winter to save some of the dryer's sensible and latent heat — using a by-pass damper to vent outdoors in summer — but the increased humidity and condensation on buffer-space glass is usually unacceptable.

If the kitchen and bathroom are vented into a buffer space, those rooms will draw replacement air from wherever they can get it most easily. A small opening connecting these rooms to the buffer space will ensure that the replacement air is from the buffer rather than coming from outdoors.

## OTHER CONSIDERATIONS
### Domestic Hot Water

Most solar domestic hot water (DHW) heating systems installed in recent years have used roof-top absorbers and an anti-freeze fluid pumped through a heat exchanger to transfer the sun's heat into the house water. This makes for a complex system with a fairly high cost.

In a buffered house, the greenhouse or sunspace provides a location for collectors where they can be protected from freezing. This allows the use of a more economical DHW system which simply preheats the water in tanks or tubes connected between the cold supply and a backup or point-of-use heater, or tubes can be connected in a gravity-flow loop to a higher storage tank.

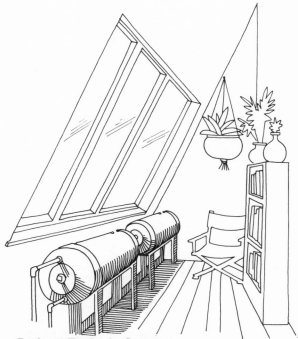

**Preheat Tanks in Sunspace**

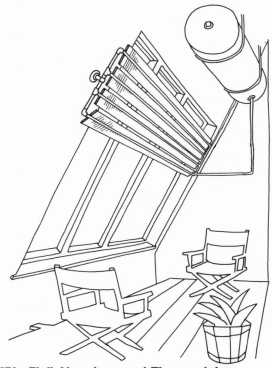

**"Big Fin" Absorbers and Thermosiphon Storage Tank**

### Monitoring Performance

The most useful measurements for the evaluation of a buffered building are of air temperatures in the shade at the sunspace (1) where coolest air enters from basement or crawlspace and (2) at the peak or warmest part, preferably with maximum-minimum thermometers. If measurements can also be taken (3) at the warm-air entrance to the crawlspace or basement, heat transfer to that space may be estimated.

Supplemental heat used may be determined by weighing firewood, and from bills for electricity or other fuels.

### Retrofits

Sunspace additions to existing houses are becoming common. In many such situations, all the heat the sunspace can produce can be absorbed in the house without overheating it, and occupants and hardy plants tolerate the night-time cooling. Better protection against night cooling may be achieved with an airflow to and from the house basement or crawlspace — if the house foundations can be insulated to reduce heat losses from the basement or crawlspace.

Insulation of existing house foundations may be accomplished from the inside, or by digging down on the outside. It is also possible on the outside to insulate by removing earth and plants to a level just below plant roots and placing a four-foot or six-foot outward-sloping skirt of rigid insulation on a carefully flattened (sand) supporting surface.

### Termites

Where termites could be a threat, a metal termite shield should cap foundation walls. Exterior insulation is also vulnerable, and should be protected by stucco extending to the footings.

Termite protection usually includes poisoning the earth around and under a building, but the poisons are unsuitable in a buffered house where the air people breathe passes across the earth surface. Reinforced concrete may be an acceptable earth-surface termite barrier.

Photo: Clarice and Dohn Kruschwitz

Photo: S J Patrick

# 6

## SUPERINSULATED HOUSES

In addition to the sun/earth buffering approach, the general approach to energy efficiency which we are finding most appreciated is superinsulation with sun-tempering. Since this approach does not usually incorporate a greenhouse or use as much glass as in a buffered design, the initial cost can be somewhat lower.

More than any other type of very efficient house, superinsulated houses can be designed in any style and retain a traditional appearance. They also need not depend on good solar access and so are suitable for virtually any site.

Most of the details of superinsulation are covered in Chapter 4, because they can be very effectively used in most types of energy-efficient houses, not just in superinsulated houses.

### Amounts of Insulation

A superinsulated house is one in which the primary design strategy is conservation, and heating needs are reduced to a tiny amount. The level of insulation and infiltration control that is necessary to keep the heating needs so small depends on the climate where the house will be located.

In our discussions of costs for a house with various insulation levels, we have called the house with around 12-inch walls "superinsulated," and have called the house with 8-inch walls just "heavily-insulated." These categories are based on those houses' heating needs in our 7500 degree-day climate.

The predicted auxiliary heat requirement of that "superinsulated" house was about 6 million Btus (6 MMBtu) per year, or about 1/2 cord of firewood. (With normal occupancy the actual heat used would probably be lower than that; the house might not be kept warm all day long every day.) The yearly heating load of the heavily-insulated house is more than twice that figure, 16 MMBtu/yr. The predicted peak hourly heating load for the "superinsulated" house is about 9000 Btu/hr, which is a little more than half the peak load for the heavily-insulated house.

In a warmer climate of about 4000-5000 degree-days, the house with 8-inch walls could possibly be considered to be superinsulated. In a very cold climate of about 10,000 degree-days, the house with 12-inch walls might not be called super-insulated.

The amount of insulation used in various parts of a house should be chosen with both a concern for the cost of insulation and an awareness of the rough proportion of heat loss from each part. (Note again the insulation levels, the heat loss proportions, and the costs in the four houses in the chart near the beginning of Chapter 4.) For example, adding insulation to an attic floor is usually inexpensive enough that it is worthwhile to insulate it more than the walls. On the other hand, if the walls are only 4 inches thick it would probably not be worthwhile to put R-60 insulation in the attic until the wall insulation was improved. In deciding the amounts of insulation in a house it is especially important to be aware of the impact of infiltration on a house's heat loss; any heavy insulation measures should always be accompanied by measures to control infiltration.

## Controlling Infiltration

Probably more than any other aspect of efficient housing, infiltration control depends on the combination of good planning by the designer and careful workmanship during construction. A study of 50 homes in upstate New York indicated that the overall tightness of a house was just as dependent on the amount of quality control during construction as it was on the specific measures used to reduce infiltration.

A "blower door" test is now available in some areas for measuring the tightness of houses. A blower door fits into an exterior doorway and has a fan which blows air either into or out of the house to pressurize it and depressurize it. At a chosen pressure difference between inside and out, the fan speed is measured to find the rate of air leakage through the building shell.

This blower door test is not yet widely used by homebuilders, but is likely to become more so soon. It is useful not only for measuring the tightness of a house, but most importantly, it also makes it easy to locate major leaks, so that corrections can be made before a house is finished. For this purpose the test should be performed after vapor and infiltration barriers are in place but before interior finish materials make them inaccessible. Preferably, the construction crew would be present during the test. The necessary equipment is too expensive for most builders to own; instead they would hire an outside firm (such as an energy-auditing firm or a company which sells insulation) to perform the test.

## South Glazing and Sun Tempering

While a superinsulated design does not require sunshine, a moderate amount of direct solar gain can be very effectively used without causing overheating/overcooling and without requiring added mass for thermal storage. This sun-tempering is achieved just by orienting the building with one wall facing within about 15 degrees of true south and then putting more windows on the south than on the other sides.

The amount of south glass which can be used in a superinsulated house without causing overheating is dependent on several variables, including the amounts of insulation and infiltration, the climate and clear sky sunshine, and the amount of thermal mass and where it is situated in the house.

Calculations based on a two-story 1500sf super-insulated house in a New England climate indicate that the south window area should be no more than 10 percent of the floor area if triple-glazing is used on the south, assuming the normal mass of frame construction and usual house furnishings.

Even less area — about 8 percent — is desirable with double-glazing, because double-glazing allows both more sun in and more heat out than triple-glazing, and so causes an increase in temperature swings, compared with triple-glazing.

(Some assumptions: 25 x 30ft outside plan measure, wood frame construction, total of 60sf of glass on E, W and N. R-30 walls and floor, R-50 ceiling, total air change rate of .25 AC/hr.)

The efficient use of added south glass represents a net energy gain over the heating season, and therefore lower heating bills. However, letting more sunshine into the house also means slightly greater temperature swings: the house will get hotter on sunny days and cool faster at night. Of course the night cooling can be prevented by using supplemental heat, but in that case the peak load on the heating system is slightly increased. A possible advantage of keeping the south window area fairly small (or else using movable insulation) is that the peak heating load can be kept to a bare minimum and the heating system be very small.

Demonstrating the effectiveness and practical achievability of superinsulation, fourteen very energy-efficient houses were built in 1980 by thirteen different builders on this block in Saskatoon, Saskatchewan. Annual heating costs were so low, and builders and buyers were so impressed, that within two years more than half of the new houses being built in the area were super-insulated.

This "Energy Showcase" demonstration was sponsored cooperatively by the Canadian and Saskatchewan governments, by the city of Saskatoon, and by the Saskatoon branch of the Housing and Urban Development Association of Canada.

## Garrison House
New Hampshire, 1982
1640 sf floor area

This is a superinsulated version of a garrison-colonial, with a full walk-out basement. There is an outstanding view to the west, so there are more windows on that side than would usually be planned; some heat conservation was willingly sacrificed for the enjoyment of the generous window area. Splayed window jambs of varnished pine provide space for plants and add beauty to the interior.

The house has a double flue chimney for a wood stove as the main auxiliary heat source and a fireplace for special occasions.

*(Designed and built by Steve Booth with Community Builders.)*

**Garrison House**

| | |
|---|---|
| 7500 | heating degree-days |
| 50 | percent of possible sunshine |
| $60,000 | cost |
| $37 | square foot cost |
| 1640 | sf house floor area |
| 85 | sf south glass area |
| .06 | ratio of south glass to floor area |
| 82 | sf glass on E, N, W |
| R-19 | floor insulation |
| R-32 | wall insulation |
| R-54 | roof insulation |
| R-12 | foundation insulation |
| high | estimated overall tightness |
| 1.8 | Btu/DD/sf |
| 18 | Million Btu's fuel consumed |
| wood | type of auxiliary heat |
| normal | estimated comfort level |

*Floor plans and sections are on a scale of 1/16" to the foot.*

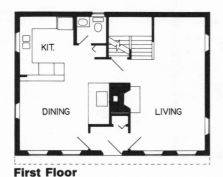

**First Floor**

**Second Floor**

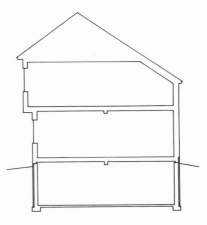

**Saltbox House**

|        |                                  |
|-------:|----------------------------------|
| **7500** | heating degree-days            |
| **50** | percent of possible sunshine     |
| **$35,000** | cost                        |
| **$30** | square foot cost                |
| **1153** | sf house floor area            |
| **85** | south glass area                 |
| **.07** | ratio of south glass to floor area |
| **38** | sf glass on E, N, W              |
| **R-19** | floor insulation               |
| **R-32** | wall insulation                |
| **R-60** | roof insulation                |
| **R-12** | foundation insulation          |
| **high** | estimated overall tightness    |
| **0.7** | Btu/DD/sf                       |
| **6** | Million Btu's fuel consumed       |
| **wd,elec** | type of auxiliary heat       |
| **normal** | estimated comfort level       |

## Saltbox House
### New Hampshire, 1982
### 1153 sf floor area

This modern salt-box was designed to provide superinsulation at a relatively low cost, with a compact and efficient floor plan. The small, one-story extension on the northeast corner adds a needed utility area and the possibility of a half bath. It also adds greatly to the exterior appearance of the house; approached from the east, the house appears much larger than it is.

The air lock entry has been considered important for energy conservation but is proving to be much less important in tight houses and could have been dispensed with. *(Community Builders design)*

**Second Floor**

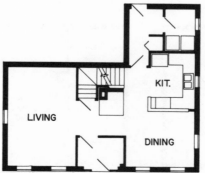

**First Floor**

1/16″ = 1ft

## Modern Cape House
Maine, 1982
1500 sf floor area

The depth of roof rafters usually does not provide enough space for thick insulation in sloped ceilings. In this modern cape, the sloping ceilings are well insulated by using a "Supertruss" (by RoKi Associates, Gorham, Maine 04038) with R-60 blown-in fiberglass insulation. The "Supertruss" design in several shapes has been developed by RoKi particularly for the superinsulation of various types of sloped ceilings, including the cape, cape with full dormer, gambrel, and saltbox.

**Modern Cape House**

| | |
|---:|---|
| 8000 | heating degree-days |
| 50 | percent of possible sunshine |
| $58,000 | cost |
| $53 | square foot cost |
| 1500 | sf house floor area |
| 53 | south glass area |
| .04 | ratio of south glass to floor area |
| 80 | sf glass on E, N, W |
| 20 | floor insulation |
| 35 | wall insulation |
| 70 | roof insulation |
| 20 | foundation insulation |
| high | estimated overall tightness |
| .55 | Btu/DD/sf |
| 6.6 | Million Btu's fuel consumed |
| elec | type of auxiliary heat |
| warm | estimated comfort level |

**First Floor**

STUDY KITCHEN LIVING DINING

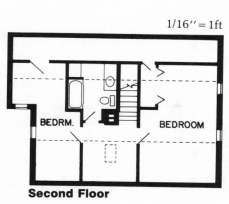

1/16″ = 1ft

**Second Floor**

BEDRM. BEDROOM

**Seacoast House**

|  | |
|---:|---|
| 7500 | heating degree-days |
| 40 | percent of possible sunshine |
| $70,000 | cost |
| 64 | square foot cost |
| 1100 | sf house floor area |
| 90 | south glass area |
| .08 | ratio of south glass to floor area |
| 12 | sf glass on E, N, W |
| 0 | floor insulation |
| R-40 | wall insulation |
| R-55 | roof insulation |
| R-25 | foundation insulation |
| high | estimated overall tightness |
| 1.45 | Btu/DD/sf |
| 12 | Million Btu's fuel consumed |
| wood | auxiliary heat |
| warm | estimated comfort level |

## Seacoast House
Maine, 1980
1100 sf floor area

The James McKenney house in Kennebunkport, which was probably the first superinsulated house in Maine, can be kept within a degree or two of its normal 72F with about a cord of wood per year. A small fan (1/70 horsepower) blows warm air from above the woodstove down to the crawlspace, which is a concrete-floored heat plenum, with floor registers to the livingspace.

Windows have sliding shutters. Roll shades on south windows can be moved from the usual position at the top of window openings to a half-height location where they can keep out unwanted spring and fall sunshine without blocking light and view at the upper glass.

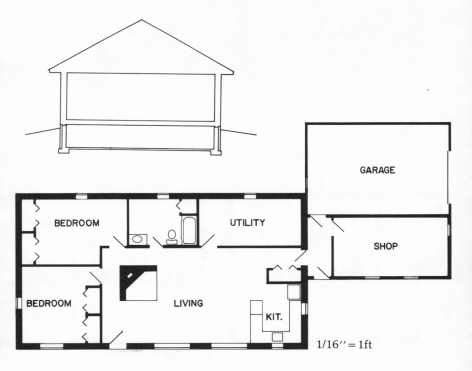

1/16″ = 1ft

## The Timberloft House
### 1510 sf floor area

The Timberloft was designed by Bob Corbett and Wally Hansen, who formerly worked with the National Center for Appropriate Technology, "to show that superinsulated houses can be an architectural delight without sacrificing energy-conserving performance." The plans are very carefully detailed so that owner-builders and contractors with no previous experience with superinsulation can use them with a minimum of difficulty.

The house is two stories on a crawl space foundation, with one bedroom and a bath on the main floor and two bedrooms and another bath on the second floor. Cathedral ceilings give a feeling of spaciousness, but the overall plan remains efficient and compact. Double stud walls and a double rafter system provide space for insulation levels of R-38 and R-50. Plumbing is economically concentrated.

The design includes an attached sunspace with water tubes for thermal storage and space for a hot tub. The sunspace would need some backup heat to prevent freezing in most cold locations. The design also includes a large air-to-air heat exchanger system, with a capacity equivalent to 1.25 AC/hr and with separate outlets for every room in the house.

The plans for the Timberloft and for two smaller houses are available from the designers; an information packet may be obtained for $2 from Corbett/Hansen, PO Box 3706, Butte, MT 59702.

### The Timberloft

| | |
|---|---|
| 7652 | heating degree-days assumed |
| 50 | percent of possible sunshine |
| $58,027 | estimated cost |
| 38 | square foot cost |
| 1510 | sf house floor area |
| 67 | south glass area |
| .04 | ratio of south glass to floor area |
| 74 | sf glass on E, N, W |
| | floor insulation |
| R-38 | wall insulation |
| R-50 | roof insulation |
| | foundation insulation |
| high | estimated overall tightness |
| 1.2 | Btu/DD/sf |
| 13.6 | Million Btu's fuel projected |
| elec | type of auxiliary heat |
| warm | estimated comfort level |

SECOND FLOOR-LOFT    1/16" = 1ft

FIRST FLOOR    NORTH

# 7

## ENERGY-CONSERVING
## CONSTRUCTION TECHNIQUES

When a carpenter looks at something needing to be done, he expects to find a straightforward way to do it, but innovative systems may have unseen traps. Looking ahead during construction is always important, but with new systems it may be especially easy to miss important preparations, and learning from experience can be costly.

How careful is careful? How tight is tight?

### SITEWORK AND FOUNDATIONS

Existing trees that are important for beauty or summer shade can be protected from damage by construction machinery by surrounding them with flag-marked stakes at a sufficient distance out from the trunk to keep the heavy machinery off the near-surface roots.

Effective and dependable groundwater drainage around footings is essential. We use 10ft lengths of rigid perforated pipe, set with a slight slope on a few inches of clean crushed stone (the 1″ size is the largest stone that can be shoveled easily). We use rigid rather than flexible drain pipe, because it is much more easily and dependably placed at a controlled slope and it will have a self-cleaning channel at the bottom if it is placed with holes equidistant from the bottom centerline.

The drain is protected from silting with a foot of crushed stone covered with building paper, or an insulation-board sloping skirt. Sand or gravel backfill material above the drainage stone allows for easy seepage of water into the stone and drain.

Footing drain pipes are carefully pitched to drain and protected with clean crushed stone.

A flush-out pipe from the drain's high point up to the surface of the ground can make possible a test of the drain, by running water from a hose through the drain pipe system to its outlet (keep the inlet capped at other times). The pipe outlet is usually at a small crushed-stone dry well, and may need to be screened to keep out small animals.

In an area with a high water table, the crawlspace inside the foundation is filled with sand to a level at least a few inches higher than that of the foundation drains. A vapor barrier near the top of the sand (and above the water level) will reduce evaporation of moisture into the crawl space.

Exterior rigid insulation is applied to the foundation just before backfilling to minimize damage to the fragile insulation boards — including the possibility of wind blowing away the insulation before it is held by the backfilling. A little insulation adhesive can be used to hold the boards in place until the fill pushes them against the wall, or the boards can be pinned with occasional nails. Backfilling requires care if the board is to remain flat and continuous against the wall. Joints should be kept tight, and a hand shovel should be used to fill against the insulation in areas where the insulation might be moved or damaged by a quick machine filling.

Insulation that would otherwise be exposed to abrasion and to ultraviolet radiation from the sky is protected with reinforced Portland cement plaster bonded to the foam. The bonding is critical; if insulation board is smooth-faced it should be scratched or roughened with a wire brush, and a chemical bonding material like Standard Drywall's Acryl-60 should be used. We extend the stucco well into the clean porous drainage backfill material, below the loam which is more likely to harbor insects that could eat into the foam.

It is very important that the finished grade be sloped away from the house to prevent surface water from running toward the foundation. Backfilling should be done carefully, to ensure even settlement. Avoid burying trash and wood scraps, which attract pests.

Reinforcing mesh is pressed into prime coating of stucco on rigid insulation.

## FRAMING
### Careful Workmanship

Awareness and care during framing can greatly reduce later problems; a carpenter understands that attention at this stage toward keeping corners plumb and walls straight will make the later installation of cabinets and other finishes much easier. Preparations are made now for many of the final details. Workers may also need to be given an understanding of the difference their care can make in the building's thermal performance and durability.

• INFILTRATION control depends on workers' awareness: With average construction, much heat is lost through tiny cracks. It is possible to be alert to these and caulk or cover them with infiltration barrier sheeting. Examples: around windows and doors, and where walls sit on floor plywood.

• VAPOR BARRIERS are a particular challenge: their continuity on the warm side of insulations requires awareness and attention. Since warm air can carry much more moisture than cool air, in winter there is a large difference in the humidity of indoor and outdoor air, and moisture from inside will be moving out wherever it can. Moisture is like heat: if there is more of it in one place than another it will migrate to equalize. A vapor barrier is to moisture what insulation is to heat: it prevents moisture from escaping from the living space (where it can contribute to occupants' wintertime health) into the outside walls (where it can condense and cause decay, and reduce the effectiveness of the insulation).

### First Steps for Tightness

Mudsills on top of a concrete foundation are usually the first framing, and require several cautions to reduce infiltration heat loss. Compressible sill-seal insulation strips are placed, without gaps, between wood and concrete. After the framing has been shimmed straight and level, cement mortar is pressed deep into the crack between wood and concrete.

Vapor barrier strips at tops of walls and at partitions will later be joined to other wall and ceiling vapor barriers to provide a continuous moisture seal.

At the start of framing work, we keep some insulation on hand to pre-insulate potentially inaccessible voids at corners, partitions, and boxed-in headers. We use a long-lasting acrylic or urethane sealant to pre-caulk potential locations of air infiltration at framing joints. Mudsills are also carefully sealed.

We have some two-foot-wide strips of thicker polyethylene on hand at the start also, for vapor barrier pre-wrapping at locations where there might be a break of vapor barrier continuity: (1) on top of the mudsills and elsewhere that floor platforms meet exterior walls (note that floor deck construction is detailed to allow for a continuous vapor barrier surface, with adequate insulation outside the vapor barrier); (2) where partitions meet exterior walls; (3) over the top of partitions of the top story where they meet insulated ceiling or roof.

## Building and Lifting Walls

Cross-hatch walls are usually fully constructed on a building floor, then tipped up into place. Plates are cut and laid out on the floor deck as in the construction of standard stud walls. Inner studs are nailed in place first, sills straightened to a chalkline or string, diagonals checked to be sure that corners are square (so they will be plumb). Braces are nailed, and horizontal girts are toe-nailed to the studs.

Accurate spacing is worthwhile, to provide for snug-fit insulation. A thickness problem can arise because the combined width of two 2x4's is 7 inches but the width of the 2x8 plates is 7-1/4 inches. When we use native rough-cut lumber for girts, we have it specially sized (to 3-3/4 inches in this case) so that the combined girt and stud thickness is equal to the 2x8 wall thickness. It is also possible to rip one edge off the 2x8 members so they will equal the combined thickness of the two standard 2x4s, or spacer strips of 1/4" plywood can be tacked or stapled across the studs under each girt.

Thick walls are heavy to tilt up but lifting jacks can make this operation easy and safe. Securing the bottom plate of the wall to the floor is important to ensure that the wall will not slide as it is lifted. This can be done by nailing short strips of sheet metal or pallet binding to sill plates and floor.

The Saskatchewan wall is also tipped up as a unit. The inner plates and studs are nailed together first, straightened and squared, and then the frame is braced with steel straps or plywood. Vapor barrier plastic is laid over the wall, with enough overlap at edges to wrap around later onto the inner face of the wall. (If plywood is used, the vapor barrier is applied before the plywood and protected by it during further construction.) The outer frame is then assembled directly above the inner, spaced from it with temporary blocks, and then fastened to the inner frame with plywood strips joining the top and bottom plates. If siding is to be applied before the wall is raised, insulation and wind barrier Tyvek are installed before the siding.

Window openings are prepared by carefully cutting the vapor barrier at sill and head and vertically at the center, then folding it around onto the inner face of the rough jambs, and adding and overlapping vapor barrier pieces at head and sill. The openings are then boxed with plywood pieces. Wall Tyvek wraps into the plywood-boxed openings, and window frames are caulked to the Tyvek at all edges.

## GLAZING

Large pieces of glass are heavy and best handled with suction cups. The edge seals of insulated units are easily damaged, and unmounted glass pieces should be stood on edge, not horizontal, on wood or other safe surface until they are installed on rubber setting blocks in their permanent locations. Glass panels need 1/4" clearance for expansion. Watch also for clearance at lag screws or other fasteners.

Rain must be kept out by the glazing system, but it is equally important to keep interior moisture vapor from penetrating the system from the inside. Glazing tapes or putty at glass should be neatly continuous, and if window frames are painted, paint should touch the glass.

In cold climates where significant condensation is expected at windows, moisture-respecting details need to be carefully carried out to prevent future damage to the windows and walls. Mildew-resistant paints or preservative stains, and durable caulking compatible with the edge seal of the insulating glass should be used in these areas.

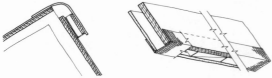

Ridge vent for unequal-pitch roofs is made from screened boxes made from plywood and strapping blocks.

## CONTROLLING AIR INFILTRATION

Careful workmanship can make a large difference in the tightness of a house (and its consequent heating costs).

To find and eliminate the tiny holes and cracks, a blower door pressurization is used by some insulation contractors, just before the gypsum is finished. Air leaks can be located with smoke or felt with the hand and corrections made.

Infiltration barriers — not vapor barriers — on the outside of insulated walls can greatly reduce air penetration and increase the effectiveness of insulations. We are currently using DuPont's Tyvek sheeting applied in full 9ft width. Tyvek is a wind barrier, but its fibrous structure permits the outward passage of moisture vapor from the insulated walls. Edges are double-lapped or sealed with caulking; tears or punctures are repaired. Tyvek is most effective if the house is completely wrapped under all exterior finish including window casings and corner boards, not just under siding.

We caulk built-up framing joints such as in built-up posts and window mullions, and where wall plates meet floors. Doors and windows are set into caulking. We normally use more than one case of one-pint tubes.

If plywood or particle board is used for sheathing, provision for venting of the insulation space should be made before applying siding. A continuous air channel may be provided inside the wall sheathing, from sill plate to the underside of roof sheathing. It may be achieved at each stud with plastic spacers which are available from some materials suppliers, or three-quarter-inch holes can be drilled in wall plates to vent each stud bay. Saw slashes can provide venting slits in the sheathing directly to the exterior if finish materials will not block the vents.

## INSULATION

If insulation is to be fully effective, it must completely fill the spaces it is intended to fill. Rigid insulation and fiberglass batts or blankets should be measured and cut accurately; insulation should be carefully pushed in — but not tightly compressed — to fill completely into corners.

Second layer of friction-fit insulation is installed on interior.

When using fiberglass, we prefer unfaced batts, even though they are somewhat less pleasant to handle than the paper-faced or foil-faced kind. They are oversized for friction fitting and generally have a higher R-value. The oversizing also makes for a better fit, and without the foil or paper covering, it is easier to see and inspect the quality of the insulation job. Compacted areas can be fluffed out and voids can be filled.

Where insulation batts are to be installed in more than a single thickness, the back (outside) layer should be fully installed first, and not covered anywhere by an inner layer of insulation until all electrical wiring or other potential disruptions to the continuity of the first, outer, layer have been completed. Stagger the joints where possible.

At ceiling or roof insulation, eave-to-ridge venting is important. Do not block the venting spaces by over-filling with insulation. In sloping roofs, where maximum insulation is desired within the framing spaces, fiber or plastic vent space strips can be stapled to the roof sheathing or framing to prevent air blockage. Provide a stop for blown-in insulation at eave cavities.

The narrow spaces around window and door frames should be filled (but not packed) with insulation, or caulked if they are too tight for insulation. This stuffing of narrow spaces is effectively done by pushing in successive layers of fiberglass with a thin stick or wood shingle tip, or by spraying in an insulating foam.

Check behind partition blocks and in exterior corners to be sure that even places that are not visible are completely filled.

Check again after electricians, plumbers, masons or others who have been working after the insulation has been installed, because they may have left some insulation out of place.

## CONTROLLING VAPOR BARRIER QUALITY

Vapor barriers should be continuous, with lapped and folded joints which will be pressed together by framing or gypsum board, and with edges caulked at windows, doors and floors.

Barrier material, if it is not placed smoothly, can bunch and make too thick a bulge under gypsum board.

Photo: International Informational Services

Blow-in insulation in process, showing baffles used to maintain venting airspace.

Vapor barrier joints are double-folded to reduce air leakage.

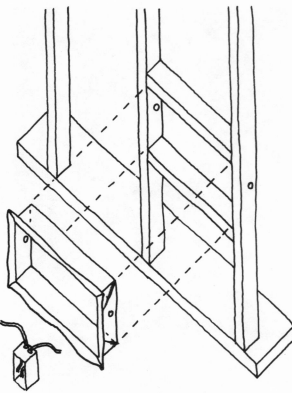

Blocking between studs provides a space for a piece of vapor barrier plastic to be folded into the wall cavity in preparation for installing wiring and an outlet box. Holes through the plastic for wiring are sealed with non-hardening accoustical sealant. Wall vapor barrier will be sealed to edges of the pre-wrapping plastic.

Where it is necessary to penetrate a vapor barrier sheet with framing, the framing can be painted with vapor barrier paint (Glidden Insul-Aid). Vapor barrier edges are taped or caulked to the framing, preferably with non-hardening accoustical sealant.

Check the quality of polyethylene film. Some may have been sitting exposed to sunlight so long that it has become brittle at the folds, and some may have been manufactured with uneven thickness. 6-mil poly is preferred over thinner material because it has better resistance to punctures and tears. Tu-Tuff is a laminated sheeting which is much stronger still.

Staple holes in the poly should be kept to the minimum necessary to hold the material in place until it is covered with gypsum board or other finish. Accidental punctures can be patched with tape. Punctures for wires and pipes should be pre-planned to have solid backing behind the poly, so they can be caulked with accoustical sealant.

Electric boxes are difficult, except with wire-chase wall systems. Wires can be brought through a separate piece of polyethylene which can be wrapped around the box and pressed against framing. It may be possible to keep boxes out of exterior walls, if the wall system does not provide for an electrical chase inside the vapor barrier. Pre-formed plastic wrappers for several sizes of electrical boxes are available from NRG Saver Distributors, Box 50, Group 32, RR1B, Winnipeg, Manitoba R3C 4A3. Surface wiring systems are also available.

Polyethylene "hat" surrounds electrical box. Wire holes are sealed, and the wall vapor barrier will be caulked to the hat.

Photo: International Information Systems

Oversee the subcontractors' work to make sure
they do not cut or drill through the vapor barrier
after it has been installed. Preplanning by designer
and owner can eliminate most changes and additions which could affect vapor barrier continuity
(e.g. extra wiring for TV, stereo, phone, intercom,
doorbells; built-ins on the outside walls).

Checklist of trouble spots:

At electrical and plumbing penetrations

At joints between wood and concrete or masonry

At any projections through the siding (e.g. decks
or balconies, air-to-air heat exchangers, or air
conditioners)

At any locations where there have been changes
that might have disturbed initially careful
construction

# 8

## EXAMPLES OF BUFFERED HOUSES

The houses described in this chapter were built in the period from 1978 to 1982. They exemplify the wealth of experimentation in design and construction that went on across the continent as people heard of the exciting performance of the early envelope homes. Eager owners and builders, many initially still uncertain about exactly what made the envelope work, developed their own designs. In spite of wide variations, the homes quite consistently worked well, and each contributed to the continuing process of discovering what is essential and what is non-essential. This diversity of experimentation has led to a better understanding of how to build houses which can be less expensive yet still very effective.

For each house we have included a statement written by someone who was closely involved with the house — owner, designer, or builder. We do not always agree with the opinions expressed in these statements.

Throughout this chapter we will use the term "envelope house" or "standard envelope house" to mean a house with buffering airspaces on the south (the greenhouse/sunspace), above most of the ceiling, in between two north walls, and below the floor (the basement or crawlspace). The east and west walls of a standard envelope house are single-layer without buffering airspaces. Many of the houses will be described with reference to this typical envelope system.

Morrison sunspace

Of the 20 houses described in this chapter, about half could be considered to be more or less standard envelope houses. Some of the variations on the basic envelope concept which are shown in this chapter are:

— Use of fans to move heated air on sunny days.

— Use of added thermal storage materials.

— Increasing the areas of buffering airspaces to buffer all four sides instead of just the north and south.

— Reducing the areas of buffering airspaces, especially on the north side and above the ceiling.

— Using some of the airflow passages for other purposes, such as for an entry or stairway.

The square foot area of inner floor which is shown for each house specifically does not include the area of the greenhouse or any other buffer or unheated portion. For virtually all owners of buffered houses, the greenhouse is an important living space (frequently the most enjoyed space in the whole house), so it may be somewhat misleading not to include it in the interior floor area, but we chose this approach so that the heating load per square foot would reflect only the portions where comfort levels are maintained. The cost per square foot is shown both with and without inclusion of the greenhouse area.

The column of data shown for each house is similar to the information given in the survey of envelope houses reported in Chapter 9. Near the beginning of that chapter there is a more complete definition of each information category.

Floor plans and sections are on a scale of 1/16″ to a foot.

*The Beale house is a two-story envelope house providing very pleasant space at a low cost. This design was also adapted for the modular home shown in the smaller photographs.*

*One economy is in the foundation, which extends only two feet below grade although frost depth can be four feet. By extending the foundation insulation in a horizontal skirt, the footing and the crawlspace earth are well protected and foundation costs were reduced. (Community Builders design)*

Galen Beale's diary told her story in "The Double Shell Solar House":

June 12 — Moved onto land and set up tent — have narrowed the house design down to an envelope — like there being no cement mass floors or walls, a greenhouse at last, lots of light in the house without having to cover the glass at night.

June 25 — I have everything I need in three apple boxes in this tent! Don doing the house drawings — I really like how he opened up the space.

## Beale House
New Hampshire, 1979
1152 sf inner floor area

### Beale House

| | |
|---:|---|
| 7500 | heating degree-days |
| 50 | percent of possible sunshine |
| $35,000 | cost (owner-contracted) |
| $30 | square foot cost (inner house only) |
| $27 | square foot cost (incl. greenhouse) |
| 1152 | sf inner house floor area |
| 144 | sf greenhouse area |
| 174 | sf vertical south glass area |
| 116 | sf sloping south glass area |
| 0.25 | ratio of south glass to inner floor |
| 83 | sf non-buffered glass area |
| 0 | sf buffered glass not on the south |
| 22 | ft envelope height |
| R-19 | ceiling insulation |
| R-19 | first floor insulation |
| R-13 | inner of double walls insulation |
| R-26 | single wall insulation |
| R-18 | outer of double walls insulation |
| R-19 | roof insulation |
| R-15 | foundation insulation |
| low | estimated overall tightness |
| 0.35 | Btu/DD/sf |
| 3.0 | Million Btu's fuel consumed |
| wood | type of auxiliary heat |
| normal | estimated comfort level |
| 100 | maximum winter greenhouse temp. |
| 36 | minimum winter greenhouse temp. |

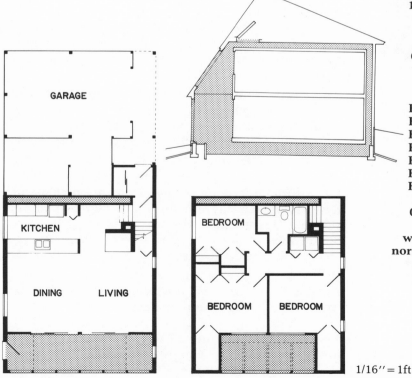

1/16″ = 1ft

**First Floor**          **Second Floor**

July 15 — Got the final drawings from Don — really don't understand them at all.

July 16 — Beginning to work on the lumber list, which makes building seem possible. Have decided to do the contracting myself as the way I can save the most money. Tried to write a schedule for myself — two days away from the house with the kids each week, one day for making baskets.

July 21 — Fascinating conference on envelope houses at [the Burns] house. It's great to be able to go over and see that house in the process of being built while I'm building mine.

July 23 — Driveway in — looks great. Foundation — don't feel elation about that — just fear. This is not going to be fun.

July 29 — Finally decided on a builder — foundation is to be poured tomorrow, then blocks — mason is skeptical about the foundation. Trying to hire as much done locally as possible — feel better now that I can see progress, but the house sure is little. Have tried to make a list of what I will worry about and what I will not. I will not worry about November, fancy food, a social life, I will worry about the kids and the cost of the house. So many decisions to make — all biggies — trying to narrow it down and eliminate some options — no plywood, plastic, metal.

August 6 — Have been hauling lumber from Minery's sawmill in the unending 90 degree heat, trying to get ahead before we start — trying to pile it logically. House construction starts tomorrow.

August 11 — Working like a dog! The lumber list was way off, given it up, just guessing now, takes a lot of time, everyone trying to figure what we need tomorrow each day but the crew is great and very relaxed. Guess this is the way it is going to be. Spent most of the very long week chasing 2x8s — struggling to get the materials here — millions of trips.

August 14 — Crew off for two days again because I can't keep enough lumber here — we're working 6 am to 6 pm which is a great way. But I really wish I could just stay home all day and work on the house and someone else would hassle with the stores. A phone today — that will make life easier. I put up a mailbox — getting territorial.

August 24 — House is coming along. I know I'm doing it the hard way. Rough lumber very uneven — takes lots of time. Shopping for everything, endless hassle about materials — don't think the crew has worked a full week yet. They are getting impatient at my disorganization — can't blame them, but I do seem determined to do it my way. Going to start the barn tomorrow to keep the crew busy while waiting for the roof trusses.

September 5 — House stalled again. Sawmill broke down. School started, the summer so short. Fear the crew will quit — they have a point. Second floor walls up. It's a nice house.

September 15 — Crew quit. Trusses cancelled because of high winds. House and barn all framed. for $12,000. I wonder who I'll get — creeping panic!

September 18 — Found another crew.

September 20 — Trusses finally came — they're wrong! Not enough and the pitch is wrong. New trusses will take another week. All stopped on the house for two weeks now, trying hard to keep remembering the things I am not going to worry about. First frost.

September 21 — Went sobbing over to Don's, told him I was beginning to take the trusses personally, he said I should read a book called "You Are Not the Target!" We ate some wild grapes and discussed the house — so calming and nice. Thought it would be ok to board in before the trusses come. Inexperience makes you overcautious and slows everything down but it's cheap.

September 27 — Getting cold — have to figure out at this point how to best spend the rest of my money — hire yet another crew to do the roof? Going to plywood the roof to save time. Moved the tents into the barn, good to get out of the wet in the morning.

September 29 — Trusses came. House looks huge now. Crew coming to do roof. Now quite used to carpenters coming and going, it's nice while they're here.

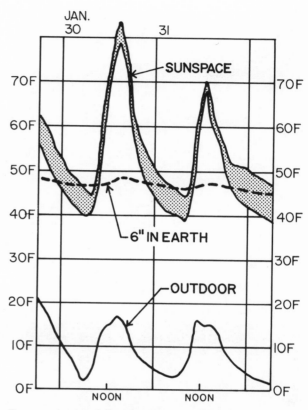

Temperatures of the greenhouse and the crawlspace earth during the last two days in January. The performance of the greenhouse is fairly typical for many envelope houses: temperatures rising to around 80F on sunny days and bottoming out somewhere around 40-45F on cold nights.

October 5 — Going to do it — move in! Stuffed the envelope with fiberglas and got a space heater. We're all delighted to be in — the kids say it's great now because there are no more worms in the tent — what little troopers they have been. It's much colder in the house than in the tent and feels so closed in — can't hear any of the night sounds.

October 17 — Snow. Don's crew coming — very professional — Don comes each morning and I'm amazed at how much of what he does that I did. Roof on, greenhouse glass finally in. Did it with the help of strong friends — the glass company brought the glass (heavy) and suction cups to hold on with — I couldn't even watch. Got away for a night but I have this anxiety if I'm gone for long that eats at me, I always want to be back here working on the house. I have not kept to my schedule of getting away with the kids, it's been all-consuming.

View through greenhouse.

October 30 — Sheetrock tomorrow. Got bricks from woods and piled them in the house to keep warm — what a mess — they are old and the mason hates them because they're so uneven.

November 13 — Sheetrockers really here AND carpenters AND a new mason!

December 14 — Taking time now to enjoy the house. Sitting, reading, watching the sunsets. Having so much glass on the south is wonderful, makes you just want to sit and look at the view without feeling you should go outside. The house is always so light — I love it. Planted the greenhouse with Boc Choy, Chinese cabbage, lettuce, spinach — it never needs watering it seems. I shoveled all the dirt for the earth beds in by hand, had a loader dump the earth at the door and now I don't see a need for the wide aisle for a wheelbarrow — takes up a lot of space.

December 22 — Going away for Christmas — it's a whole new way of thinking to just walk away from the house and not have to find someone to come every day to stoke the woodstove. Still lots of leaks in the house, needs caulking, but it's so pleasant to live in with the high humidity and so much glass that a day can never look dreary because you are always looking through the greenhouse and this incredible green color of whatever is growing there. I wish there were a window on the north so I would have some warning when people come in the driveway because it is very hard to hear with the house this tight — and a sliding door instead of a window on the second floor because it's so much warmer in the upper greenhouse it's a great place for hanging plants, I am continually climbing out the window. I love heating my house with the sun and the house just doing it by itself — no machines to fiddle with, no rushing around at night to button up the house. To wake up in the winter with the ice on the windows and to feel the sun come up and eat away at the ice and turn it into pools of water in the greenhouse and then gradually raise the temperature of the house — it's a wonderfully natural process.

West Virginia modular house was adapted from Beale plans.

Factory-built sections were assembled at site.

## Morrison House
New Hampshire, 1981
2600 sf inner floor area

*The Morrison house is a fairly large envelope house that was designed and built by the owner. The rooms on the east and west ends are normally closed off and unheated, providing additional buffering for the central living area.*

Comments by Jake Morrison, owner and builder:

I had always wanted to build a house, and since this was my first attempt, simplicity and conventional methods were essential both in design and construction. I endeavored to use the lessons and styles of the colonials, updated with modern materials and design to develop a basic "hearthside" lifestyle with one large area of kitchen/living function built around a fireplace/woodstove. Function and simplicity prevailed with a basic square containing bedrooms upstairs, the plumbing stacked on the north wall, and all wrapped in the envelope. Building covenants required that the house be at least 2600 square feet, so I again imitated a New England farmhouse by surrounding the hearth with unheated rooms that provide a thermal buffer on the east and west walls.

Siting of the house was simple: true south provided a perfect view of a small pond. The land contours encouraged an earth berm on the north and a ground level basement with acrylic panels on the south. Low vertical glass allows some direct gain on the earth store underneath the house.

All foundation drains were doubled and connected to a plenum to act as an air pre-conditioner, but it

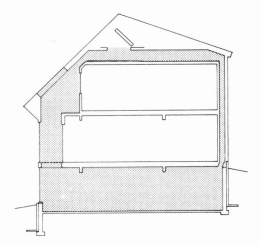

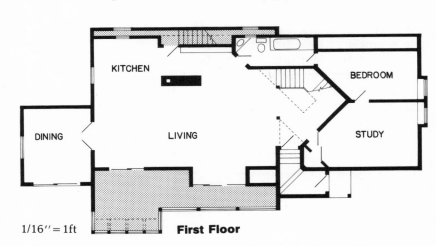

1/16″ = 1ft          **First Floor**

KITCHEN

BEDROOM

DINING

LIVING

STUDY

was discovered that the system could not provide the volume of air necessary to be effective for hot weather cooling.

The dimensions and design of the envelope airspaces were "state of the art" at the time. A one-foot airspace was used on the north but I believe that less would work, as mine is just about full of plumbing and yet a reasonably vigorous air flow can be detected on sunny days. I installed the stairs from the garage to the back entrance in the envelope, and this seems to be a good use for part of the buffer space.

The glazing area appears to have been a bit too generous, with 600 square feet. The three 42sf skylights have leaked ever since they were installed, cause summer overheating and are impossible to clean. I made worse mistakes or miscalculations in other areas and have been constantly amazed at both the summer and winter performance, which I believe speaks well for the forgiving nature of the envelope design.

The outer walls of the house are 6" thick, and the inner house is normal 2x4 construction. My guess is that this arrangement is just backwards — put the best thermal membrane around the heated living space. The outer wall is primarily a low-stress thermal buffer.

Building an envelope house seemed relatively easy. Despite many frustrations and challenges, I don't feel that the envelope house is any more difficult or expensive to build than any conventional home.

**Morrison House**

| | |
|---|---|
| 7500 | heating degree-days |
| 50 | percent of possible sunshine |
| $90,000 | cost (owner-built) |
| $34 | square foot cost (inner house only) |
| $32 | square foot cost (incl. greenhouse) |
| 2600 | sf inner house floor area |
| 256 | sf greenhouse area |
| 332 | sf vertical south glass area |
| 126 | sf sloping south glass area |
| 0.18 | ratio of south glass to inner floor |
| 40 | sf non-buffered glass area |
| 0 | sf buffered glass not on the south |
| 26 | ft envelope height |
| R-19 | ceiling insulation |
| R-19 | first floor insulation |
| R-13 | inner of double walls insulation |
| R-19 | single wall insulation |
| R-19 | outer of double walls insulation |
| R-30 | roof insulation |
| R-10 | foundation insulation |
| low | estimated overall tightness |
| 0.51 | Btu/DD/sf |
| 10.0 | Million Btu's fuel consumed |
| elec | type of auxiliary heat |
| normal | estimated comfort level |
| 86 | maximum winter greenhouse temp. |
| 38 | minimum winter greenhouse temp. |

**Second Floor**

## Alpha House
Ontario, 1980
2110 sf inner floor area

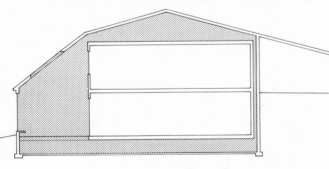

1/16″ = 1ft

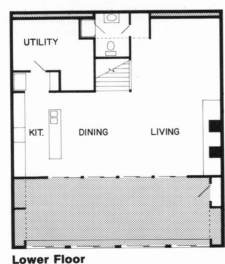

**Lower Floor**

UTILITY

KIT.   DINING   LIVING

*The Alpha House is a two-story envelope house which is set into a south-facing slope. It is located in Ontario, Canada, in a climate with 7200 degree-days and only 35 percent sunshine. (The percent sunshine is the average amount of possible winter sun for a location. Around 50 percent is more typical, and of course the amount of sunshine available has a significant effect on the performance of any solar house.) The house has a thermostatically-controlled fan at the top of the sunspace to encourage airflow around the envelope loop.*

Comments by John Hix, architect:

I see two valid approaches to energy-efficiency in houses. One is to build homes very tightly with high insulation standards and small glass areas (Saskatchewan house). The second is to use similar specifications but with larger glass areas and storage mass. I became fascinated with the double envelope concept of Lee Butler linking a sun room solar collector space with the earth beneath a house for storage. The attraction was at first economic bacause thermal mass, so necessary in a passive solar home, is usually expensive to introduce into conventional lightweight timber construction prevalent in Canada.

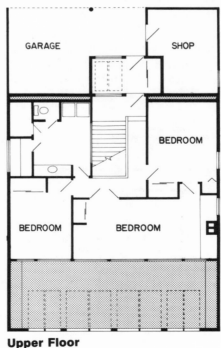

GARAGE      SHOP

BEDROOM

BEDROOM      BEDROOM

**Upper Floor**

The aesthetic qualities of the double envelope sun room combined with its speculated thermal performance caused me to organize the construction of the double envelope "Alpha" house with a major addition to the Butler design, that of adding a fan in the attic to assure a good thermal link between the sun room and the crawlspace earth below. By doing this, the controversy over whether or not the air moves by natural convection in a double envelope home in the collection mode could be set aside. The aesthetic attraction of the double envelope home, which provided energy efficiency with large glass areas, was considered a more attractive solution than a small-windowed design. The results of experience with Alpha house — its economic construction, its aesthetics, its comfort conditions and low heating bill — all support the attraction of the hybrid double envelope concept in Canada.

The house has worked beyond our expectations. We are particularly pleased the way the earth below tempers the sun room space and we are able to keep house plants healthy throughout the winter with no heat.

The compact plan is of special benefit to the heat balance. As an architect, it was tempting to make my own home rather expressionistic. I am pleased that the house is simple and restrained. That was my best design decision.

Performance of the house can be illustrated by looking at its behaviour on the coldest days of the first year's heating season. The gas furnace was never used between 11 p.m. and 8 a.m. All temperatures were taken at 8 a.m.

|  | Living | Sunroom | Outdoors |
|---|---|---|---|
| December 24 | 54F | 36F | 24F |
| December 25 | 56F | 28F | -18F |
| December 26 | 60F | 32F | 14F |
| December 27 | 60F | 38F | 18F |
| December 28 | 62F | 40F | 20F |
| December 29 | 64F | 43F | 30F |
| January 1 | 54F | 32F | 8F |
| January 2 | 54F | 28F | -18F |
| January 3 | 54F | 28F | -18F |
| January 4 | 50F | 28F | -30F |
| January 5 | 54F | 36F | -2F |

(In the second year, the sunroom never went below 34F, with no heat on in the house. We believe that the earth temperature rises over time.)

**Alpha House**

| | |
|---|---|
| 7200 | heating degree-days |
| 35 | percent of possible sunshine |
| $65,000 | cost |
| $30 | square foot cost (inner house only) |
| $26 | square foot cost (incl. greenhouse) |
| 2110 | sf inner house floor area |
| 432 | sf greenhouse area |
| 200 | sf vertical south glass area |
| 200 | sf sloping south glass area |
| 0.19 | ratio of south glass to inner floor |
| | sf non-buffered glass area |
| | sf buffered glass not on the south |
| 26 | ft envelope height |
| R-28 | ceiling insulation |
| R-28 | first floor insulation |
| R-11 | inner of double walls insulation |
| R-28 | single wall insulation |
| R-33 | outer of double walls insulation |
| R-28 | roof insulation |
| R-15 | foundation insulation |
| | estimated overall tightness |
| 0.68 | Btu/DD/sf |
| 10.3 | Million Btu's fuel consumed |
| gas, wood | type of auxiliary heat |
| | estimated comfort level |
| | maximum winter greenhouse temp. |
| 28 | minimum winter greenhouse temp. |

## Julian House
New Hampshire, 1981
2460 sf inner floor area

*The original envelope concept has been modified considerably in this design by Hank Huber. Airspaces in the north wall and above the ceiling have been eliminated entirely. A heat storage system has been added between the floor of the main living space and the insulated basement ceiling. A fan is used to blow hot air from the top of the sunspace down through a small duct into the floor storage system, which contains 850 water-filled plastic jugs suspended between the 16" floor trusses. The air circulates around the jugs and out into the north side of the basement, and then back up into the sunspace.*

Comments by Hank Huber, designer:

In early 1980 I was hired by William and Sydna Julian to design an energy efficient family residence. We set out to plan an air envelope home similar in concept to others I had done in the southern New Hampshire area. Preliminary cost estimates led us away from the full double shell toward a superinsulated single shell structure with a moderately sized earth-buffered sunspace.

Since the Julians wanted a wood floor throughout and a full basement, I decided to store the surplus high-grade heat absorbed in the sunspace in a thermal mass located just below the first floor space. Fan convection moves air through this

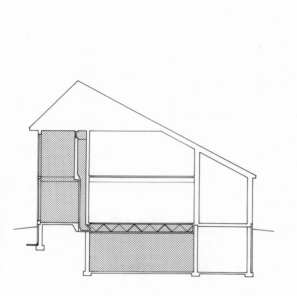

**Lower Floor**

subfloor mass, comprised of water jugs and the truss floor frame itself, promoting absorption of the higher-grade (70F — 80F) component of the warm sunspace air. Heat is released into the living space by direct radiation and convection. The lower-grade (60F — 70F) component of the circulating air is absorbed into the basement thermal mass as the air finds its way back into the sunspace. Minimum sunspace temperatures are moderated by the low-grade heat provided by the rather large basement thermal mass. The scheme may be described as a "bi-level" heat storage system.

Passive solar tempering is provided by both direct and indirect south glazing. A masonry floor is incorporated into the family dining area (southeast corner) for additional thermal mass. Insulating shades are planned for the large direct-gain apertures but are not yet installed.

Conservation measures on the single shell building start with 2″ of extruded polystyrene foam around the foundation perimeter. The earth connection is maintained by eliminating insulation below the basement slab, but a 6 mil polyethylene vapor barrier was installed under the slab for moisture control. A typical exterior wall was framed with 2x6 at 2′ on center (R-19 fiberglass) with 1″ polyisocyanurate insulation applied to the inside of

**Julian House**

| | |
|---|---|
| 7000 | heating degree-days |
| 50 | percent of possible sunshine |
| $135,000 | cost |
| $55 | square foot cost (inner house only) |
| $51 | square foot cost (incl. greenhouse) |
| 2460 | sf inner house floor area |
| 192 | sf greenhouse area |
| 336 | sf vertical south glass area |
| 90 | sf sloping south glass area |
| 0.17 | ratio of south glass to inner floor |
| 111 | sf non-buffered glass area |
| | sf buffered glass not on the south |
| | ft envelope height |
| R-40/50 | ceiling insulation |
| | first floor insulation |
| | inner of double walls insulation |
| R-30 | single wall insulation |
| | outer of double walls insulation |
| R-45 | roof insulation |
| R-11 | foundation insulation |
| high | estimated overall tightness |
| 3.6 | Btu/DD/sf |
| 20 | Million Btu's fuel consumed |
| wood | type of auxiliary heat |
| normal | estimated comfort level |
| 90F | maximum winter greenhouse temp. |
| 40F | minimum winter greenhouse temp. |

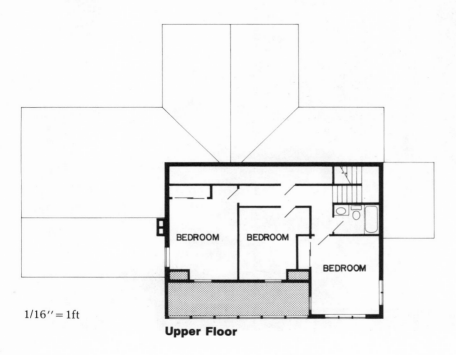

1/16″ = 1ft

**Upper Floor**

the frame. This insulation was continuous, as was an additional 6 mil vapor barrier, through the intersections with interior partitions. All non-south windows were triple-glazed. The double-glazed two-story sunspace has a net aperture of 336sf. There are 111sf of double-glazed unbuffered direct-gain windows. A 90sf active domestic hot water collector is located on the south roof.

Two high-performance axial fans circulate 700cfm of 80F — 90F air from the upper sunspace through the trusses of the first floor, which contains 850 one-gallon polyethylene water jugs. The total heat storage capacity is 7,400 Btu/degF, including the trusses and subfloor. A serpentine flow pattern ensures adequate heat transfer. A layer of 1-1/2" foil-faced polyisocyanurate insulation (R-12), nailed to the bottom of the trusses, contains the warm air and reflects radiant energy up toward the floor. Circulating air finally drops into the basement to find its way back into the sunspace.

Initial monitoring reveals temperature fluctuations from 64F — 78F, with a gain of 7F — 8F on an average sunny winter day. About 70 percent of the circulated heat gets stored in the subfloor system, with 30 percent delivered to the basement thermal mass. A greater number of small storage containers improves heat transfer into storage.

Despite minimal plant growth in the sunspace, and 650sf of living space for each of the four occupants, the tight house averages 45 percent relative humidity in mid-winter. Ducting has been provided for future installation of an air-to-air heat exchanger.

The house has been occupied since June, 1981, but the storage mass and subfloor insulation were not installed until January, 1982. Several major non-buffered glass areas will be fitted with insulating shutters to reduce yearly auxiliary heat requirements to one cord of wood or less, burned in the living room's airtight stove.

The Julian house is the first home using the "bi-level" heat storage concept. A later house built near Concord, New Hampshire achieves a higher subfloor storage temperature (around 110F) by incorporating a solar air collector in series with the sunspace. Warm sunspace air passes through the integrated site-built south roof collector prior to circulating through the subfloor heat storage system.

## Torii House
Massachusetts, 1981
2514 sf inner floor area

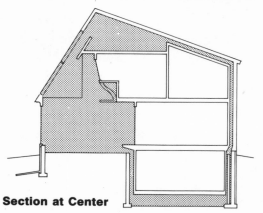

**Section at Center**

*This is a two-story envelope house which has an extra concrete block wall between the insulated north walls, for additional heat storage. Fourteen very small fans are used to blow warm air down through and around the concrete blocks on sunny days. A striking feature of this house is the Japanese garden-style greenhouse, which is set into the house so that the rooms to the east and west of it have a view onto it from the side. Torii House is located in Leverett, MA.*

Comments by William Starkweather, designer and builder:

Wisely or unwisely, I built this house as much as possible from raw materials. The only thing "hired done" by others was the digging of the hole. I used the USDA formula for block-bond and found that approach to masonry wall construction both easy to execute and pleasurable. We enjoyed the ringing sound of several hammers aligning the stacked blocks (pennies turned out to be the most available and reasonably-priced shims!). Painted block-bond walls are attractive without further treatment, as well as waterproof.

The 27ft by 8ft "heat wall" standing within two-thirds of the north wall is designed as a quick heat-sponge to hold a day's solar heat and give it back during the evening. With block cores aligned and 3" spaces between the block wall and the insulated foundation and inner 2x4 walls, all of the block surfaces are exposed to the air pulled through them by the fan system. Upward reverse flow is generated without fans by the heat released from the blocks, also driven by falling cooler air behind the south glass.

1/16" = 1ft

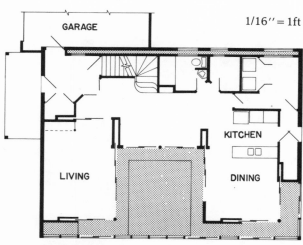

**First Floor**

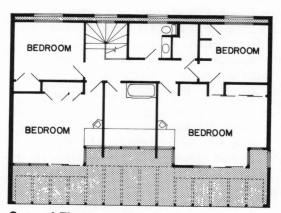

**Second Floor**

The fan system uses fourteen rheostat-controlled five-watt $10 fan units anchored to the footing and not touching (or vibrating) the wood frame. For the energy purist, this 70-watt load could be driven by solar cells. Once our computer monitoring system is finished, we are going to experiment with running the fans all the time, with the idea of completely balancing the temperatures for a more uniform envelope microclimate.

The three 12″ turbine vents in the roof proved inadequate for summer cooling, but are good for spring and fall excess heat release. Large screened louvers were added east/west for summer venting. Use of the house in the first full summer also suggested the addition of a large screened summer porch on the east, affording more living connection with the outdoors.

In subsequent designs, I plan to supply more reflected light into garden spaces (from shutterable 60-degree clerestory-like glass). Most of the available and desirable tropical plants, such as philodendrons, palms, monstera — almost everything but bamboo and schefflera — dislike full sun.

The Japanese veranda idea was well-suited to an envelope. With inner sliding glass walls that disappear into pockets, the veranda/envelope spaces become extensions of inner rooms. Even when the inner glass walls are in place, the visual connection extends the feeling of apparent space for the rooms adjoining garden and verandas.

"Heat wall" under construction inside north foundation.

### Torii House

|  |  |
|---:|---|
| 6000 | heating degree-days |
| 45 | percent of possible sunshine |
| $130,000 | cost (owner-built) |
| $51 | square foot cost (inner house only) |
| $46 | square foot cost (incl. greenhouse) |
| 2514 | sf inner house floor area |
| 308 | sf greenhouse area |
| 239 | sf vertical south glass area |
| 226 | sf sloping south glass area |
| 0.22 | ratio of south glass to inner floor |
| 75 | sf non-buffered glass area |
| 58 | sf buffered glass not on the south |
| 34 | ft envelope height |
| R-19 | ceiling insulation |
| R-19 | first floor insulation |
| R-11 | inner of double walls insulation |
| R-32 | single wall insulation |
| R-22 | outer of double walls insulation |
| R-19 | roof insulation |
| R-16 | foundation insulation |
| medium | estimated overall tightness |
| 0.40 | Btu/DD/sf |
| 6.0 | Million Btu's fuel consumed |
| wood | type of auxiliary heat |
| occas | estimated comfort level |
| 85 | maximum winter greenhouse temp. |
| 33 | minimum winter greenhouse temp. |

## "minergy" House
Massachusetts, 1980
2240 sf inner floor area

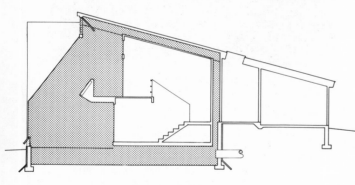

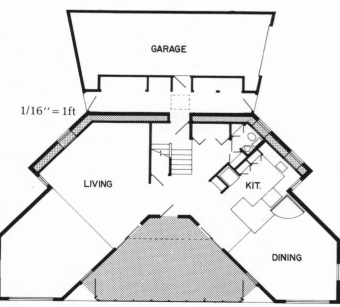

1/16" = 1ft

GARAGE

LIVING

KIT.

DINING

**First Floor**

*This house combines the envelope concept in the center portion with direct gain in the east and west wings. Thermal mass in the wings is provided by concrete slab floors and extra layers of gypsum wallboard. The two-story wings are higher than the greenhouse and block much of the eastern and western sun in the summer which could otherwise cause greenhouse overheating (especially with the large sloped glazing area). In the winter the sun's path stays more to the south and the greenhouse is not shaded by the wings.*

Comments by Douglas Holmes, owner:

"minergy house" was designed by William Mead, AIA, WM Design Group, Center Harbor, NH.

In our initial discussions with Bill Mead, we outlined to him our concept of the "minergy house" principle: MINimum enERGY requirements by design. We desired more than just a "solar heated" house. Our wish was to formulate a strategy of careful, total, integrated design to strenuously reduce the very need for energy.

Bill Mead's response to our requests was the "minergy house" design. The basic layout and form of the house, he says, were inspired by three different American designs: the Anasazis' Pueblo Bonito, Frank Lloyd Wright's Solar Hemicycle, and the Balcomb house in Santa Fe. The design demonstrates economy, through careful choice of materials and methods and clever use of all the space; simplicity, through use of native woods which are oiled or varnished, not painted, and pigmented concrete floors, rather than tiles, and several other such concepts that Wright used; and conservation, through clever detailing of insulation and weather-stripping, double-glazing facing SE, S and SW and triple-glazing in all other windows, low-flow faucets and shower heads, dual plumbing drains to permit gray water recycling with heat recovery, and careful choice of all appliances.

The sunspace was constructed with recycled steel framing and cypress glazing bars, around 60 years old. Only the glazing and aluminum flashing are new. The highest walls are sun-facing; the roof planes all drop off toward the north and the cold winter winds. The garage and mudroom, like the shed or storage room of an old New England saltbox, protect the north side of the living space from winter storms.

The chevron-shaped floor plan opens up to the sun during the heating season, while the second floor rooms which surround the sunspace provide extensive shading during the summer. All of the living spaces on the lower floor open on to the sunspace, which is not only the most delightful space in the house, but is the primary source of heat, and will become a significant source of food as well. The air heated in the sunspace proceeds around a convective loop surrounding the core of the house. The end rooms, both upstairs and down, have pigmented concrete floors and large south-facing windows. Two 40ft long "earth tubes" for cooling the air in the sunspace and convective loop are connected to the crawlspace walls on the NE and NW sides of the house.

We did most of the solar engineering calculations and analysis ourselves. Throughout the design phase, together with our architect, we visited solar homes in various parts of the U.S., read nearly all of the available books and articles, attended National Passive Solar Conferences and talked or corresponded with many owners, builders, and architects whose experience we felt might be pertinent to our project.

Initially, Bill proposed a massive wall between the sunspace and the rest of the house. After reading Michael Phillips' article in the Summer, 1978 issue of Co-Evolution Quarterly, however, we decided to incorporate a convective loop rather than a mass wall.

Heat load calculations and the choice of window sizes proceeded fairly smoothly. We were less certain, however, about the appropriate amounts of mass and the placement of it, so as to properly absorb the incoming radiation and keep the room temperature within the comfort range. Our design was tested against every bit of guidance or rule-of-thumb we came across. In the direct gain areas, we felt that 6" of concrete downstairs was a good choice with slab-on-grade construction. Upstairs, we compromised our desire for 4" with about 2-1/2" poured over the beam-and-plank ceilings below, together with three layers of 5/8" gypsum wallboard on the ceilings and northerly walls in the end rooms. In the core areas, surrounded by the convective loop, we felt reasonably sure there was already sufficient mass in the joists, studs,

## "minergy" House

| | |
|---|---|
| 6500 | heating degree-days |
| 55 | percent of possible sunshine |
| $122,000 | cost |
| $54 | square foot cost (inner house only) |
| $48 | square foot cost (incl. greenhouse) |
| 2240 | sf inner house floor area |
| 284 | sf greenhouse area |
| 144 | sf vertical south glass area |
| 356 | sf sloping south glass area |
| 0.22 | ratio of south glass to inner floor |
| 200 | sf non-buffered glass area |
| 30 | sf buffered glass not on the south |
| 27 | ft envelope height |
| R-16 | ceiling insulation |
| R-11 | first floor insulation |
| R-11 | inner of double walls insulation |
| R-30 | single wall insulation |
| R-30 | outer of double walls insulation |
| R-19 | roof insulation |
| R-10 | foundation insulation |
| high | estimated overall tightness |
| 0.62 | Btu/DD/sf |
| 9.0 | Million Btu's fuel consumed |
| wood | type of auxiliary heat |
| normal | estimated comfort level |
| 95 | maximum winter greenhouse temp. |
| 40 | minimum winter greenhouse temp. |

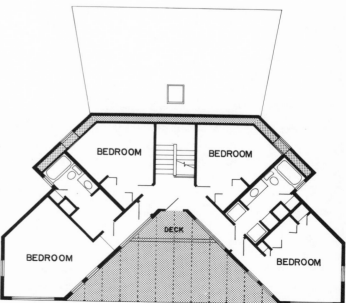

**Second Floor**

Inside north airflow space, showing foil-faced insulation taped to edges of penetrating beam.

wallboard, plaster, bathroom tiles and all the other conventional materials, so that we did not add any additional mass in these rooms.

The outer walls have an R-value of about 30 and the upper ceilings are about R-45. The vapor barrier was very carefully detailed and equally carefully installed. For a barrier we used the aluminum foil facing of Thermax board, together with aluminum foil tape at all the joints between sheets of Thermax, as well as around the very few penetrations, such as the main electrical conduit, and the 6x12" ceiling beams downstairs.

During the final design work, we had extensive conversations about all the "what if" situations we could think of: what if we experienced some form of indoor air pollution? (we provided for, but did not install, an air-to-air heat exchanger), what if the temperature swings in the end bedrooms are excessive? (provisions were made in detailing window frames to allow the installation of thermal shutters), what if someone does finally come onto the market with a "heat mirror" material to raise the R-value of double-glazed windows from about 2 to perhaps 4 or 5? (we selected south-facing windows which have a removable inner glazing panel, so that such heat mirror film could be inserted later). Our desire in these matters was to consider all of these eventualities during the design so that we would not make choices which would cause undue difficulties in the future, if and when we elected to make changes or additions.

During our first year in "minergy house" we were frequently asked if we would do things differently if we were to do it all over again. Even today we can still answer such a question negatively. Basically, the house has met our expectations and fulfilled our desires. On the other hand, the thermal performance of the house is not perfect (whatever that would be), and for some people might in fact be termed unsatisfactory. For example, the so-called comfort region is often described as being between 65F and 80F; by this criteria, the living spaces in "minergy house" are not always within these limits. Last summer, when we had several days of afternoon temperatures in the upper nineties, the end bedrooms reached temperatures of around 88F. However, the shape and form of this

Jonathan Goell

Domestic hot water panels are in front of the greenhouse balcony at upper right

house is such that it ventilates very well, even without strong breezes, so that at night the bedrooms cooled off quickly to comfortable temperatures.

In the middle of winter, temperatures downstairs in the early morning are commonly in the 58F to 61F range. It should be noted that we have never attempted to load either of the woodstoves so that it would burn throughout the night. We generally let the fires burn out about the time we go to bed. Thus, we check the weather on such chilly mornings: if it's going to be sunny, we know that the temperatures will reach the mid 60's by 10 AM or so, but if it is raining or snowing, or overcast, and we're going to be in the house for a while, such as on weekends, then we start a fire in one or both of the stoves. The house is so well insulated that a brief fire, of say 10 lbs. of wood, will quickly raise the temperature by 7-10F. In other words, the stoves do not really heat the house, in the sense that a conventional furnace does, they just serve as chill chasers. If these excursions outside of the comfort range should ever prove unnacceptable to us, or to future owners, it is now obvious that small electric baseboard units could easily maintain the desired temperatures at very little cost. Wiring for such units has already been installed in the walls.

In summary, "minergy house" represents a design combining various passive solar systems, balanced with extensive conservation measures. The estimated total annual load is about 95 million Btu. The sun provides about 64 million Btu, the appliances, lighting and occupants provide about 20 million Btu and the two small woodstoves provide the other 11 million Btu, consuming less than one cord of wood to do so.

Hourly temperatures recorded (by hand!) at three locations at the "minergy" house. Temperatures rise rapidly when the sun shines into the low-mass greenhouse, and then with the sunshine gone they fall to a base temperature around 50F.

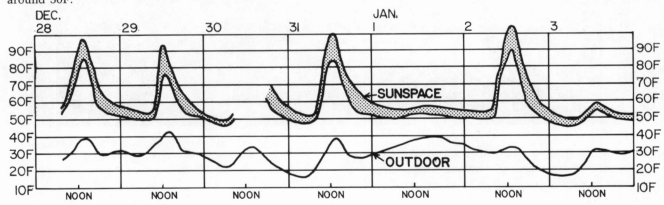

*The Williams house is similar to other envelope houses except that it does not have a separate attic plenum — the airflow goes through openings near the greenhouse ceiling into the second story bedrooms and through the bedrooms into the north wall cavity. There is also added thermal storage in the crawlspace, in the form of 500 gallons of water.*

*With the lack of a complete thermal barrier between second story rooms and the buffering spaces, it would be expected that the temperature of those rooms would closely follow the temperatures of the buffer spaces (especially the greenhouse), and become quite cold when those spaces cooled down on winter nights. Therefore the monitoring results (see chart) are surprising since they show the rooms maintaining quite comfortable temperatures even when the connected airspaces are cold. It seems clear that some heat must be lost from the second-story rooms at those places, but both the daily and the seasonal performance do not seem to be affected much by the additional loss.*

**Williams House**
Colorado, 1980
1655 sf inner floor area

Comments by Max Williams, designer, builder, and owner:

The main purpose for building this home was to show people that an energy-efficient home could be designed and built for a cost which was within the market price for medium-income people. For this reason the home is a basic box design using conventional construction methods. The total cost was about $75,000, excluding the land. Not including the garage, this was about $35 per square foot and was well within the reach of most middle-income people in this locale in 1980.

To help control the costs we eliminated the double roof system. This did two things. One, it eliminated the doubling effect of the cost of labor and materials for the second roof necessary in the complete double envelope. Secondly, it eliminated all the problems of compliance to local building codes relative to the airspace or chase. The decision to make the upper loop air move through the living space does not seem to have lessened the dynamics of the air flow, as this house performs as well as, if not better than, other envelope homes in the area. But obviously there are more factors involved than just design shape that affect the air flow.

**Fire Safety Caution**
Where airflow goes through sleeping areas, careful fire prevention and detection measures are particularly important since smoke and heat can travel rapidly from one room to another.

Northwest stairwell corner, showing top of north inner wall ending below second floor ceiling.

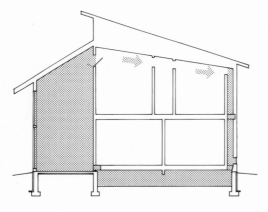

As they lose heat they draw warmer air toward them at the top, where the openings to the chase wall are located.

After living in the house for one full year we have noticed several things. There is an obvious connection between the soil temperature in the crawlspace and the air temperature in the lower part of the sunspace. Especially at temperatures below 60 degrees in the sunspace the two temperatures will fluctuate alike. We are now collecting data from the temperature monitoring system that should substantiate this. The interior living space has fluctuated between 62F and 75F during the past year. The sunspace fluctuates between 48F and 85F.

There is an internal loop connecting the two levels of the home. The stairway and heat chases at the woodstove comprise the connecting channels. This enables the two levels of the house to maintain about a one-degree spread in temperature. This uniformity is noted throughout the year.

Summer cooling is accomplished by controlling solar gain with the overhang at the sunspace and operating certain windows. The vents along the north wall proved to be unnecessary and all but two have been removed. Cooling does not present a large load in our area and can usually be easily handled.

The peak or high point of the loop is at the top of the sunspace. This affords the loop air a reservoir of warmed air to draw from as the loop air movement begins to slow in the early evening. The

1/16″ = 1ft

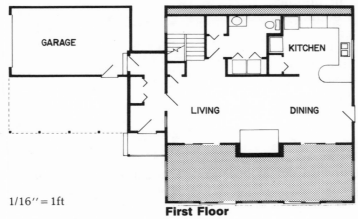

**First Floor**

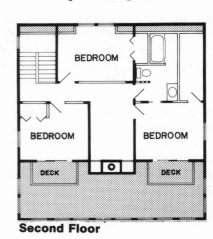

**Second Floor**

thermal masses in the sunspace radiate heat in the evening to help the loop air to continue to circulate until late in the evenings. The best method I have used so far to observe this phenomenon is by the use of smoke sticks. Primitive, but does indicate air movement if used with care. The second floor ceiling slopes toward the north chase wall to guide the air somewhat into the north chase. The windows in the north bedroom act as magnets to the air as they are not insulated inside the north chase wall.

The woodstove is needed only when outdoor temperatures fall below 20 degrees or during periods of two or more days without sun. The wall behind the woodstove is a direct thermal mass that absorbs both the sun's heat and the heat from the stove.

Hourly temperatures recorded at five locations at the Williams House. At the times when auxiliary heat is used to keep the living room warm, such as on the afternoon and night of December 27, the temperature of the sunspace peak follows the living room temperature fairly closely, because of the direct opening between upstairs rooms and the sunspace. At those times it is surprising to see how mucher colder the lower portion of the sunspace is; the sunspace air is apparently very stratified. In the period shown here, the sunspace only got as cold as 45F, when the outdoor temperature was around zero. The temperature in the earth under the house stayed around 52 to 54F. The presence of sun shining into the sunspace is indicated on the graph by the sunspace temperature line going above the other temperature lines.

### Williams House

| | |
|---|---|
| 5800 | heating degree-days |
| 60 | percent of possible sunshine |
| $75,000 | cost |
| $45 | square foot cost (inner house only) |
| $35 | square foot cost (incl. greenhouse) |
| 1655 | sf inner house floor area |
| 320 | sf greenhouse area |
| 262 | sf vertical south glass area |
| 0 | sf sloping south glass area |
| 0.16 | ratio of south glass to inner floor |
| 64 | sf non-buffered glass area |
| 66 | sf buffered glass not on the south |
| 22 | ft envelope height |
| R-38 | ceiling insulation |
| | first floor insulation |
| R-15 | inner of double walls insulation |
| R-20 | single wall insulation |
| R-19 | outer of double walls insulation |
| R-38 | roof insulation |
| | foundation insulation |
| medium | estimated overall tightness |
| 0.42 | Btu/DD/sf |
| 4.0 | Million Btu's fuel consumed |
| wood | type of auxiliary heat |
| normal | estimated comfort level |
| 86 | maximum winter greenhouse temp. |
| 48 | minimum winter greenhouse temp. |

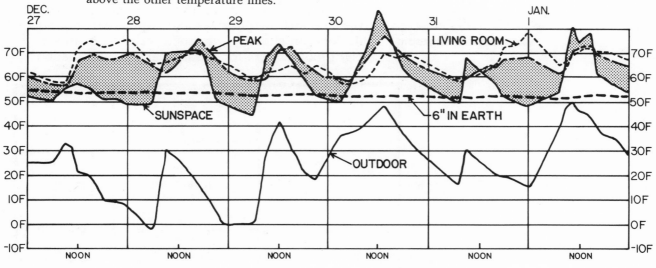

## Burns House
New Hampshire, 1979
1100 sf inner floor area

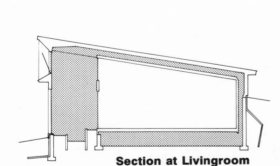

**Section at Livingroom**

1/16″ = 1ft

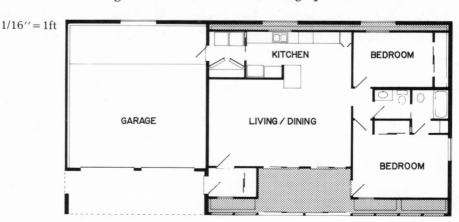

GARAGE

KITCHEN

BEDROOM

LIVING / DINING

BEDROOM

*This house has been described in many publications, because it was one of the earlier envelope houses and probably the first to be extensively monitored. It is a standard envelope design, with all of the south glazing installed vertically. The high south glazed vents throw light far back into the living room. (Designed and built by Community Builders.)*

The Burnses wanted a one-story retirement home that would require little maintenance or purchased heat. They assumed that the house would be set into the earth on their south-sloping site and use as much sunshine as the New Hampshire weather allowed. They wanted good growing space for plants and a Clivus Multrum composting toilet.

We had been studying the thermal envelope concept and had received an Appropriate Technology grant under which we proposed to build and evaluate a buffered house; the Burnses agreed that their house could be part of that project.

We were not certain that a single-story house would have enough envelope airflow height for a thermosyphon gravity flow to work, so we prepared ourselves for the possibility that we might be building just a well insulated — and buffered — passive solar house.

An efficient framing system evolved from a combination of factors: a four-foot glass module, a long span across the open living space, maximum use of the strength of standard lumber sizes, and most of all a desire to avoid running framing members through airflow passages and vapor barriers. We decided on site-built plywood-web box beams on eight-foot centers for the long spans, and made the

**Burns House**

| | |
|---:|:---|
| 7700 | heating degree-days |
| 50 | percent of possible sunshine |
| $70,000 | cost |
| $63 | square foot cost (inner house only) |
| $54 | square foot cost (incl. greenhouse) |
| 1100 | sf inner house floor area |
| 200 | sf greenhouse area |
| 350 | sf vertical south glass area |
| 0 | sf sloping south glass area |
| 0.32 | ratio of south glass to inner floor |
| 8 | sf non-buffered glass area |
| 30 | sf buffered glass not on the south |
| 23 | ft envelope height |
| R-19 | ceiling insulation |
| R-13 | first floor insulation |
| R-13 | inner of double walls insulation |
| R-22 | single wall insulation |
| R-22 | outer of double walls insulation |
| R-19 | roof insulation |
| R-12 | foundation insulation |
| high | estimated overall tightness |
| 0.28 | Btu/DD/sf |
| 2.4 | Million Btu's fuel consumed |
| wood | type of auxiliary heat |
| normal | estimated comfort level |
| 90 | maximum winter greenhouse temp. |
| 40 | minimum winter greenhouse temp. |

living space width three times and the bedrooms twice this eight-foot module. Ceiling, roof and floor purlins between the beams are 2x6s.

For summer venting we planned high south hinged windows above the greenhouse. They are seldom opened. Light through these high windows reaches far back into the living spaces.

At this time we thought we could best provide natural growing conditions in the greenhouse growing beds by keeping them directly connected to the

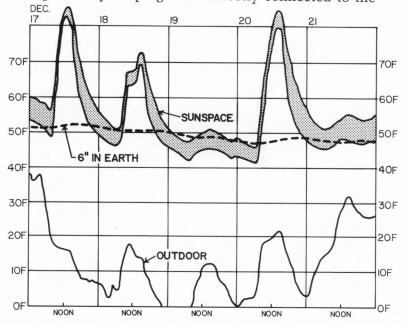

Temperatures at the Burns House for five days in December. Sunspace temperatures are shown for both the peak (upper line) and lower portion (lower line). During the cold and cloudy weather of the 19th, the temperature of the earth is falling, but the earth still provides enough heat to keep the sunspace above 42F. During the winter the earth at a six-inch depth dropped from an October high of 56F to a January low of 46F. The lowest greenhouse temperature for the winter was 40F.

**Burns Greenhouse**

deeper earth under the building, so air from the crawlspace and basement enters the greenhouse from the east and west ends.

A basement was not needed except for Clivus Multrum space, but when we later began to plan the measuring of temperatures under and around the building we were glad we could observe the thermal performance of both crawlspace and basement.

Thermal mass was added at the foundation walls on the north by placing the vertical exterior insulation about three feet out from the concrete wall, thus bringing the three feet of earth into closer thermal contact with the wall. Horizontal insulation a foot below the earth surface connects with the upper wall's insulation and caps the additional earth mass. We have since judged that the value of the additional mass did not justify such a difficult placement of insulation.

**Fawcett Livingroom**

*This traditional-appearance New England gambrel has no greenhouse or sunspace, only a 12" airspace between inner and outer glazing on the south. There is no air passage at all on the north, and instead there are airshafts at east and west ends, connecting the attic space to the crawlspace, and buffering the east and west windows. The ratio of south glass area to floor area is small compared with most envelope houses. Located at 15 Trotting Park Rd, Lowell, MA 01854 (Community Builders design).*

When we planned this no-greenhouse house, we wanted an exterior that would seem normal and usual in New England, and that could provide a lot of living space for as little cost as possible. We thought that buffering the east and west windows with airflow channels between attic and basement might accomplish more than a north-wall airspace.

The 12" airspace between inner and outer south glass allows more direct solar gain in the living space than is possible with deeper sunspaces, but air circulation through the house and basement keeps interior temperatures moderate. We had thought it might be necessary to use a sun-absorbing tinted surface on the inner south glass, to stop the sun's direct heat within the buffer space.

The house was built at the upper edge of a steep north slope, with a walk-out basement. Little winter sun could reach the south side of the house during the first year because it was shaded by tall pines of a state forest. Auxiliary fuel usage was about two cords of mixed hard and soft wood that year, but came down to one cord after the pines were cut and night shutters were installed on the south glass.

## Fawcett House
### Massachusetts, 1980
### 1972 sf inner floor area

**Fawcett House**

| | |
|---:|:---|
| 5600 | heating degree-days |
| 55 | percent of possible sunshine |
| $53,000 | cost (owner-built) |
| $26 | square foot cost (inner house only) |
| $26 | square foot cost (incl. greenhouse) |
| 1972 | sf inner house floor area |
| 0 | sf greenhouse area |
| 120 | sf vertical south glass area |
| 0 | sf sloping south glass area |
| 0.12 | ratio of south glass to inner floor |
| 0 | sf non-buffered glass area |
| 101 | sf buffered glass not on the south |
| 31 | ft envelope height |
| R-20 | ceiling insulation |
| R-20 | first floor insulation |
| R-13 | inner of double walls insulation |
| R-20 | single wall insulation |
| R-13 | outer of double walls insulation |
| R-20 | roof insulation |
| R-10 | foundation insulation |
| | estimated overall tightness |
| 1.09 | Btu/DD/sf |
| 12.0 | Million Btu's fuel consumed |
| wood | type of auxiliary heat |
| normal | estimated comfort level |

**First Floor**

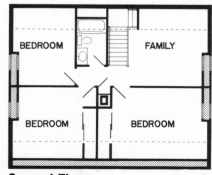

**Second Floor**        1/16″ = 1ft

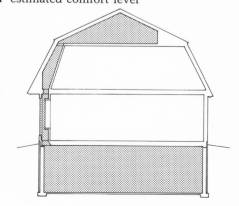

## Studio House
New Hampshire, 1981
1600 sf inner floor area

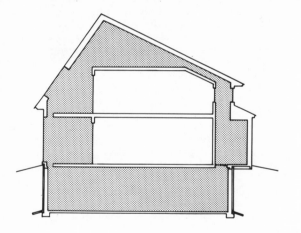

*The double north wall system of a standard envelope house has been eliminated in this design, and instead there is a nearly square 24sf air passage on the north, which goes down through the principal entryway. (Community Builders design)*

In this house for two artists, a principal requirement was a lot of light, so we used more glass on the east, west, and north than we normally might.

We had expected to buffer the large north windows, but budget limits turned us to reduce the north airflow to just the entranceway. We have been glad we did. Even though the north windows were not buffered, and the living room during the first winter was connected to the greenhouse because the chimney brickwork was incomplete, supplementary wood heat requirements the first year were little more than a cord.

This was our first use of a simpler, reduced-surface north airspace. Gravity airflow has not been measured but seems superb, and sunspace and attic air temperatures never get higher than 85F. Two of the sloping glass panels are intended to cover a domestic hot water heater, so the greenhouse gets a little more overhead light and heat than it will when the DHW absorbers are installed.

1/16″ = 1ft

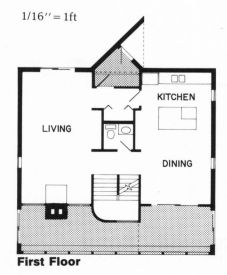

**First Floor**

LIVING

KITCHEN

DINING

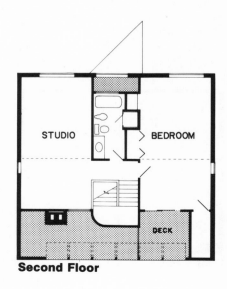

**Second Floor**

STUDIO

BEDROOM

DECK

Views through sunspace from diningroom and bedroom.

## Studio House

| | |
|---:|:---|
| **7500** | heating degree-days |
| **50** | percent of possible sunshine |
| **$75,000** | cost |
| **$47** | square foot cost (inner house only) |
| **$41** | square foot cost (incl. greenhouse) |
| **1600** | sf inner house floor area |
| **224** | sf greenhouse area |
| **224** | sf vertical south glass area |
| **136** | sf sloping south glass area |
| **22.5** | ratio of south glass to inner floor |
| **132** | sf non-buffered glass area |
| **84** | sf buffered glass not on the south |
| **33** | ft envelope height |
| **R-26** | ceiling insulation |
| **R-13** | first floor insulation |
| **R-13** | inner of double walls insulation |
| **R-26** | single wall insulation |
| **R-26** | outer of double walls insulation |
| **R-38** | roof insulation |
| **R-12** | foundation insulation |
| **high** | estimated overall tightness |
| **1.17** | Btu/DD/sf |
| **14** | Million Btu's fuel consumed |
| **wood** | type of auxiliary heat |
| **normal** | estimated comfort level |
| **90** | maximum winter greenhouse temp. |
| **32** | minimum winter greenhouse temp. |

Daily maximum and minimum temperatures of four locations for the month of January. The upper single lines show the range of daily temperatures at the greenhouse peak, while the box-bars show the temperature range close to the bottom of the glass. The temperatures recorded near the glass are about 15 to 20F cooler than the temperatures for the greenhouse peak. Outdoor temperatures are in the lower part of the chart. The horizontal dashed line shows the earth temperature at a six-inch depth.

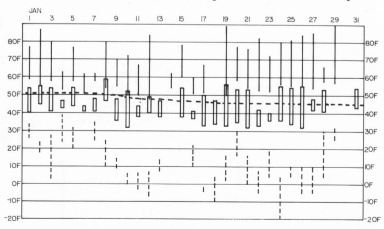

## Winnisquam House
New Hampshire, 1981
1120 sf inner floor area

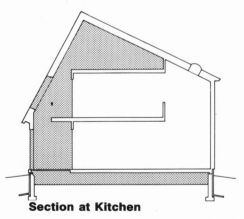

**Section at Kitchen**

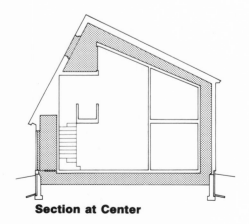

**Section at Center**

*This house is an envelope design, but it has only an eight-foot long double wall and airspace on the north wall, rather than one which extends across the full length of the house as in a typical envelope house. This house is also notable for its two-story interior spaces (cathedral ceiling) and good natural daylighting. (Community Builders design)*

Natural daylighting from high-ceiling spaces has been much enjoyed in this house. Most of the high light comes through buffered south glass, but there is a skylight in the kitchen — which also adds light to one side of a second-floor study.

"A two-way fireplace, some two-story spaces, and lots of light" were the owner's first requirements. The result is a visually pleasing interior, but the two-way glass-door fireplace could not deliver even the little heat that it was intended to deliver. The owner eventually decided to close in the dining-room side with brick and install a steel fireplace insert facing the livingroom.

The owner often travels and is usually away daytimes, so without supplementary heat the livingspace can be down to the fifties Fahrenheit when she returns. The fireplace fire warms the house too slowly, so radiant electric units are turned on for quicker comfort.

Summer comfort is easily maintained by opening windows at night, usually keeping them closed during hot days.

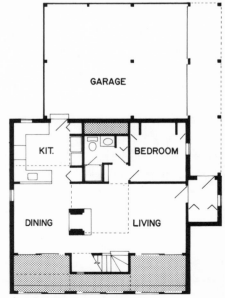

1/16″ = 1ft

**First Floor**

## Winnisquam House

| | |
|---:|---|
| 7500 | heating degree-days |
| 50 | percent of possible sunshine |
| $65,000 | cost |
| $58 | square foot cost (inner house only) |
| $51 | square foot cost (incl. greenhouse) |
| 1120 | sf inner house floor area |
| 157 | sf greenhouse area |
| 156 | sf vertical south glass area |
| 120 | sf sloping south glass area |
| 0.25 | ratio of south glass to inner floor |
| 70 | sf non-buffered glass area |
| 8 | sf buffered glass not on the south |
| 30 | ft envelope height |
| R-26 | ceiling insulation |
| R-19 | first floor insulation |
| R-13 | inner of double walls insulation |
| R-26 | single wall insulation |
| R-19 | outer of double walls insulation |
| R-38 | roof insulation |
| R-11 | foundation insulation |
| high | estimated overall tightness |
| | Btu/DD/sf |
| | Million Btu's fuel consumed |
| wd,elec | type of auxiliary heat |
| cool | estimated comfort level |
| | maximum winter greenhouse temp. |
| | minimum winter greenhouse temp. |

**Second Floor**

## Rich House
New Hampshire, 1981
1470 sf inner floor space

*The most distinctive feature of this modified envelope house is the location of the primary greenhouse: in the attic. There is a smaller sunspace in a more conventional first floor location, used also as an entryway. Many envelope house owners have found their attics useful for storage space, for guest sleeping, or as a snug refuge on sunny winter days, but we found no other house using the attic as a greenhouse.*

*The air passage from attic to crawlspace is through a north entryway/light shaft, under a skylight. The airspace on much of the south side is only 12" wide to allow more inner living space. A fan located at the attic ridge is used to increase air circulation around the loop and reduce attic overheating. (Community Builders design)*

The attic greenhouse design was inspired by seeing slides of Bengt Warne's three-story glass-walled "Nature House" in Sweden. The explanation accompanying the slides spoke of each successivly higher level being suited to plants with higher temperature needs. The scenes of people sitting in their sunny plant-filled attic retreat on a snowy winter day were irresistable.

Some practical reasons for this attic greenhouse design were:

— The space was there and available, already insulated and lined with gypsum (for fire protection) as an air flow space.

— The sloping attic roof provided an opportunity for good overhead lighting for plants without additional construction cost other than for installing glass in place of roofing.

1/16″ = 1ft

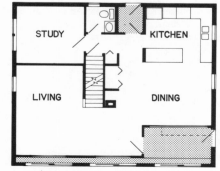

**First Floor**

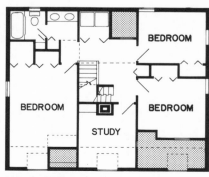

**Second Floor**

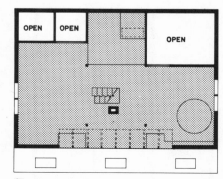

**Attic**

— Attic temperatures tend to be higher than first floor temperatures, and can provide conditions for warmer climate plants to grow well.

— The privacy of the attic location gives a special "hide-away" or retreat character to the space.

The owner does indeed enjoy spending time in the attic greenhouse retreat. A hot tub is planned. Space has been left for a dumb-waiter elevator to carry plant trays between the lower sunroom/entry and the attic.

Another unusual feature of this house is the stairway used for access to the attic. Each tread is half-width, alternating left and right so the stairway uses only about half as much floor space as conventional stairs.

Attic greenhouse in "Nature House."

**Rich House**

|         |                                        |
|--------:|----------------------------------------|
| 7500    | heating degree-days                    |
| 50      | percent of possible sunshine           |
| $70,000 | cost                                   |
| 47      | square foot cost (inner house only)    |
| 44      | square foot cost (incl. greenhouse)    |
| 1470    | sf inner house floor area              |
| 90      | sf greenhouse area                     |
| 156     | sf vertical south glass area           |
| 32      | sf sloping south glass area            |
| 0.13    | ratio of south glass to inner floor    |
| 120     | sf non-buffered glass area             |
| 2       | sf buffered glass not on the south     |
| 33      | ft envelope height                     |
| R-19    | ceiling insulation                     |
| R-19    | first floor insulation                 |
| R-13    | inner of double walls insulation       |
| R-26    | single wall insulation                 |
| R-19    | outer of double walls insulation       |
| R-38    | roof insulation                        |
| R-13    | foundation insulation                  |
| high    | estimated overall tightness            |
| 1.36    | Btu/DD/sf                              |
| 15.0    | Million Btu's fuel consumed            |
| wood    | type of auxiliary heat                 |
| high    | estimated comfort level                |
|         | maximum winter greenhouse temp.        |

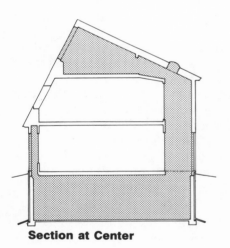

**Section at Center**

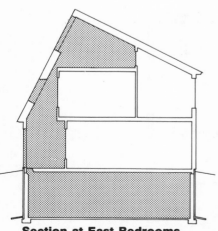

**Section at East Bedrooms**

## Holdos House
Pennsylvania, 1980
2600 sf inner floor area

*This house has double walls on all four sides, with airspaces on the east and west in addition to the north and south as in a normal envelope design. A fan is used to encourage air circulation. Summer cooling is accomplished by blowing warm air from the top of the second story into the basement.*

Comments by Bill Brodhead, designer:

In my work on envelope houses I have tried to design different types and styles so that we could evaluate the performances. As we progressed I began to realize that the real benefit of the envelope was not as a heat-replacing solar system but rather as a solar-and-earth-tempered insulation system. The real benefit is its ability to reduce heat loss to a minimum even at glass areas.

While we were designing John and Bette Holdos' house it only made sense to envelope the whole building. We used a 6″ cavity on the north to allow enough room to provide some natural convection if the power were to fail. On the east and west we used only a 3″ cavity. When we monitored the different sides later on we found no difference in temperature. All the windows in the house were doubled using a triple-glazed casement on the outside and a single sheet of 1/4″ glass that slides on a track on the inside. Since that time I've used only thermopane units on the inside; these can also be made to operate on sliding tracks.

To insure good heat transfer downward we incorporated a 10″ squirrel cage blower with a 1/4 hp

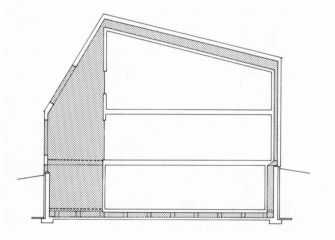

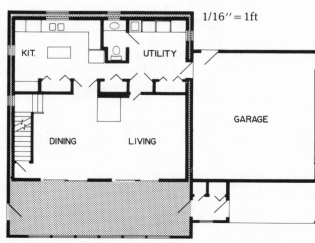

1/16″ = 1ft

**First Floor**

KIT.

UTILITY

DINING    LIVING

GARAGE

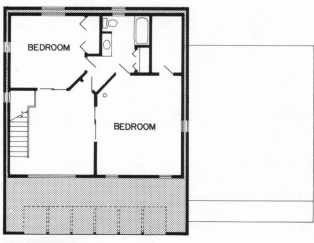

**Second Floor**

BEDROOM

BEDROOM

motor wired to an attic thermostat. We blocked off a portion of the envelope cavity to create a small plenum air chase from the peak to the basement. This low-cost item helped to maintain 50-55F foundation temperatures all winter. The basement also contains five rows of 12″ cinder blocks turned on their side with a wooden deck built on top of them. The air from the fan passes through the block cores to help increase heat transfer down.

The house sits in sparse tree cover and during the first winter I was surprised to see that peak greenhouse temperatures were only in the 70's. With the help of a pyrometer, we discovered that there was 50% loss of light transmission through the tree branch cover. This means we could have eliminated 50% of the south glass, cut down the trees and achieved the same or better performance! The fuel requirement was approximately 3/4 of a cord of wood.

Another interesting event was the summer cooling. We had incorporated three 12″ roof turbines and two 60ft long 12″ cooling tubes on the north. Unfortunately there was a difficult set of plugs to remove in the peak of the greenhouse to open the turbines and the owner failed to do it. All summer, whenever the greenhouse overheated, the winter fan would start up and pump the heat down into the basement. By the end of July the basement foundation walls had hit a peak temperature of only 66F and in August it had fallen back down to 62F. This goes to show that there is much more thermal mass in the basement than we could ever use. Since that time we have been using the basement as a closed loop cooling system. We bring warm air directly from the peak of the second floor down into the basement and return it up the stairs to the main floor.

We also had some problems with condensation on the glass the first winter and retrofitted the greenhouse with an air-to-air heat exchanger. This seemed to cure the problem very quickly.

**Holdos House**

| | |
|---|---|
| 6000 | heating degree-days |
| 40 | percent of possible sunshine |
| $80,000 | cost |
| $30 | square foot cost (inner house only) |
| $28 | square foot cost (incl. greenhouse) |
| 2600 | sf inner house floor area |
| 300 | sf greenhouse area |
| 159 | sf vertical south glass area |
| 165 | sf sloping south glass area |
| 0.12 | ratio of south glass to inner floor |
| 32 | sf non-buffered glass area |
| 36 | sf buffered glass not on the south |
| 24 | ft envelope height |
| R-14 | ceiling insulation |
| R-0 | first floor insulation |
| R-11 | inner of double walls insulation single wall insulation |
| R-16 | outer of double walls insulation |
| R-24 | roof insulation |
| R-14 | foundation insulation |
| high | estimated overall tightness |
| 0.83 | Btu/DD/sf |
| 13.0 | Million Btu's fuel consumed |
| wood | type of auxiliary heat estimated comfort level |
| 71 | maximum winter greenhouse temp. |
| 34 | minimum winter greenhouse temp. |

## Martin House
Tennessee, 1980
864 sf inner floor area

*The Martin house is an envelope design, but from the street it appears to be a traditional log cabin. The builders reported no special difficulties or extra costs resulting from the combination of envelope design with log cabin construction.*

Comments by Howard Switzer, designer and builder:

Erelene Martin asked us to build her a traditional log cabin in spring 1979, and we asked her if we could build her a solar house since the cost of energy was going up. She said she wouldn't mind at all, being on a limited income, but asked if we could still make it look like a traditional log cabin. We told her we thought we could, and since she wanted the inside sheetrocked anyway we saw it as a perfect opportunity to build a double envelope.

Having just heard about the success of the double envelope design in other parts of the country we were quite confident and enthusiastic about building one. Erelene loved the idea of having a greenhouse all along the south side of the house, so with the cedar logs she had been saving for years we began construction the following winter.

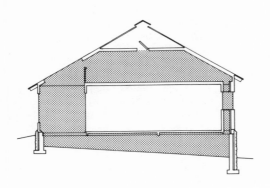

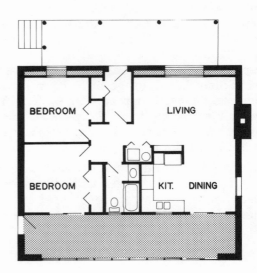

1/16″ = 1ft

We built it on a cost-plus basis, and after it was all done she said her 34ft x 38ft log cabin cost her $42,000 or $32.50 per square foot. She says she couldn't be happier with its performance, having burned "something less than a cord" of wood her first winter. She especially enjoys demonstrating its cool interior to visitors on hot sunny Tennessee summer afternoons.

### Martin House

| | |
|---|---|
| 3500 | heating degree-days |
| 45 | percent of possible sunshine |
| $42,000 | cost |
| $48 | square foot cost (inner house only) |
| $36 | square foot cost (incl. greenhouse) |
| 864 | sf inner house floor area |
| 288 | sf greenhouse area |
| 228 | sf vertical south glass area |
| 240 | sf sloping south glass area |
| 0.54 | ratio of south glass to inner floor |
| 30 | sf non-buffered glass area |
| 44 | sf buffered glass not on the south |
| 18 | ft envelope height |
| R-19 | ceiling insulation |
| R-11 | first floor insulation |
| R-20 | inner of double walls insulation |
| R-13 | single wall insulation |
| R-20 | outer of double walls insulation |
| R-11 | roof insulation |
| R-19 | foundation insulation |
| | estimated overall tightness |
| 3.97 | Btu/DD/sf |
| 12.0 | Million Btu's fuel consumed |
| wood | type of auxiliary heat |
| | estimated comfort level |
| | maximum winter greenhouse temp. |
| | minimum winter greenhouse temp. |

## Clayton House
Ohio, 1981
1872 sf inner floor area

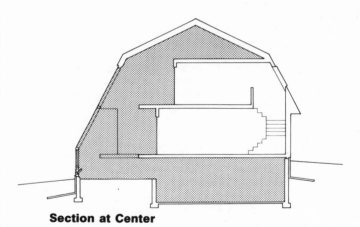

**Section at Center**

*In this two-story gambrel, there is no separate north airspace at all, and air circulates by natural convection on sunny days through the interior of the house. Heated air from the greenhouse enters the upstairs through large windows and sliding glass doors, and then moves through the stair opening to the first floor and back into the greenhouse. There is a fan and duct at the center of the house, but they are used only in the fall, to provide extra heat to the earth mass of the basement in preparation for winter. The house is very well insulated and tightly constructed. Located near Lancaster, OH; for further information call evenings (614) 969-4353.*

Comments by Doug Clayton, designer and builder:

I had been working with the "envelope" concept for about a year and a half when it finally came time (after years of planning and site searching) for me to help my folks design and build their new home. Both the "envelope" and "super-insulation" concepts appealed to me and I conceived of this house as a hybrid of the two approaches to low energy dwelling.

Rather than provide a complete envelope around the inner dwelling I chose to eliminate the north plenum and complete the convection loop (on sunny winter days) by circulating the greenhouse-warmed air through the super-insulated house interior via large operable window areas. The windows between the greenhouse and inner house provide more vent area than a 1ft by 36ft north

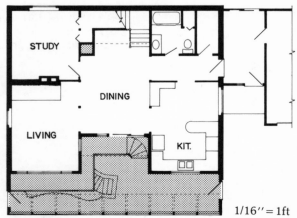

**First Floor**

STUDY

DINING

LIVING

KIT.

1/16″ = 1ft

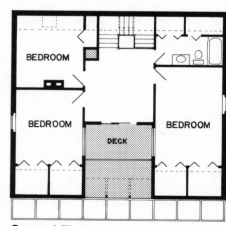

**Second Floor**

BEDROOM

BEDROOM

BEDROOM

DECK

plenum, and expose the convection currents to much more thermal mass than is available in a north plenum. I think this strategy stores the solar gain where it is most useful and most easily conserved.

A thermostatically controlled fan/duct system provides a means of moving warm air from the envelope peak directly to the basement so that it is possible to begin charging up the earth mass beneath the building in the early fall with Btu's that would normally be wasted by venting. The fan and duct are not used in winter.

The house is bermed into a southeast-facing slope and the south aperture is shifted down so that direct solar gain is provided onto the earth and concrete masses of the greenhouse and basement areas. I hoped that by applying heat at the bottom of the convection loop more lift would be provided and the convective flow rate would be increased. Coupled with the relatively small aperture area (compared to many early envelopes) the direct gain onto the massive portions of the building and the natural convection enhancement configuration resulted in a temperature behaviour that is more uniform throughout the buffer zone and more stable over time.

Large louvered vent areas (over 50sf each top and bottom) with manually operated doors are provided for silent summer cooling.

The sloping glass of the greenhouse is 1″ tempered thermopane units. In an attempt to save money, the vertical glazing was accomplished by installing two sheets of regular double strength glass in each opening. This was NOT worth the additional labor cost and fogging annoyance.

To minimize winter heat loss and summer heat gain, exterior walls are of 10-1/2″ fiberglass-filled double stud construction, and exterior windows (except on the south) are triple glazed. Interior walls facing the greenhouse are 5-1/2″ fiberglass-filled and interior windows are double glazed. This interior glass is shaded from direct solar gains in the summer by ceiling and deck overhangs.

Great care was taken with the 6-mil poly vapor barrier. All seams were rolled and/or taped and electrical boxes on the outside walls were carefully sealed to the vapor barrier. A 50cfm Mitsubishi heat exchanger provides fresh air.

**Clayton House**

| | |
|---:|---|
| 5500 | heating degree-days |
| 35 | percent of possible sunshine |
| $84,000 | cost (owner-built) |
| $44 | square foot cost (inner house only) |
| $38 | square foot cost (incl. greenhouse) |
| 1872 | sf inner house floor area |
| 336 | sf greenhouse area |
| 106 | sf vertical south glass area |
| 352 | sf sloping south glass area |
| 0.24 | ratio of south glass to inner floor |
| 75 | sf non-buffered glass area |
| 0 | sf buffered glass not on the south |
| 32 | ft envelope height |
| R-27 | ceiling insulation |
| R-19 | first floor insulation |
| R-19 | inner of double walls insulation |
| R-32 | single wall insulation |
| R-19 | outer of double walls insulation |
| R-30 | roof insulation |
| R-13 | foundation insulation |
| high | estimated overall tightness |
| 0.78 | Btu/DD/sf |
| 8.00 | Million Btu's fuel consumed |
| wood | type of auxiliary heat |
| normal | estimated comfort level |
| 80 | maximum winter greenhouse temp. |
| 45 | minimum winter greenhouse temp. |

Large openings between the greenhouse and basement facilitate air circulation.

The foundation is insulated with 3″ of extruded polystyrene extending down three feet and then out horizontally 5 feet in an attempt to include more earth mass within the basement area for seasonal heat storage.

A simple "cubic" exterior shape was chosen to minimize the ratio of surface area to volume and thus the heat loss and costs. The basement and attic make useful work and storage spaces of the buffered zone. The basement is bright and cheery with its full width view into the greenhouse and beyond. Visual complexity is provided in the greenhouse interior by varying the depth and height and providing many floor levels with decks and terracing.

Sloping glass was used primarily because the greenhouse was envisioned as a serious food production environment that needed year-round light. With its earth floor the greenhouse is simply an enclosed garden. Growing directly in the earth provides the plants' roots with a more stable temperature and moisture environment. The sloping glazing also provides year-round gain to the dual tank "bread box" domestic hot water heater that is located within the greenhouse where it is protected from freezing.

The sloping glass of the greenhouse seemed to extend naturally into a gambrel roof shape which matched nicely many old barns in the neighborhood.

Auxiliary heat is provided with two small airtight wood stoves, one located in the inner house and the other in the basement. My folks have chosen to manage the house in a way that I never dreamed of. The basement/greenhouse is maintained above 65F throughout the winter by burning about 2 cords of wood. They enjoy the sewing, laundry and shop areas of the basement right through the winter without wearing the sweaters that I imagined would be necessary. Tomatoes set fruit and ripen in December. Even on the frequent overcast days of the central Ohio winter the greenhouse may be open to the inner house and the second floor sitting deck is enjoyed almost daily.

With this VERY mild climate surrounding about half the surface area of the inner house, only an occasional fire is needed there, consuming perhaps 1/4 cord of wood per winter.

On sunny winter days the greenhouse reaches peak temperatures in the upper 70's by noon and remains steady through the afternoon. Internal natural convection breezes are frequently seen fluttering the leaves of the plants, and when you stand in an open doorway to the greenhouse on a sunny day you can feel the air rushing past into the inner house.

The inner house is in the low 70's with 50 to 60 percent relative humidity all winter long, and the temperature variation throughout the inner house is never more than a degree or two F.

During summer nights air is drawn through the house by the greenhouse's stack effect creating an artificial breeze even on warm still nights. Then in the morning the house is buttoned up, staying cool all day. 76F is the highest temperature yet observed in the interior.

Now if only there were an easy access to the outside clothesline. If only the driveway wasn't so steep; if only there was a good root cellar. If only...

Daily maximum and minimum temperatures in the greenhouse during the month of January. (The single lines in the upper part of the chart show the range of temperatures at the greenhouse peak, and the box-bars show the temperatures at the basement-greenhouse opening. Outdoor temperatures are at the lower part of the chart. The horizontal dashed line is the earth temperature at a depth of six inches.) The greenhouse is often heated by the basement woodstove, but in the period of January 24-28 the stove was not used at all, and with outside temperatures dropping below zero the greenhouse stayed above 50F.

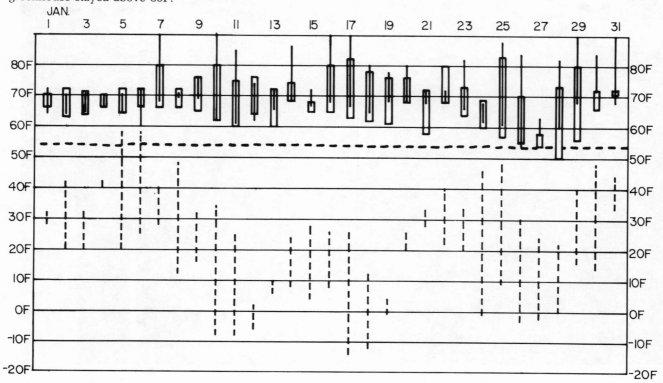

*This buffered house applies so many innovative concepts it required a book for itself: "Saunders' Shrewsbury House", by William A Shurcliff. Norman B. Saunders, its principal designer, is an engineer, inventor and solar consultant living in Weston, MA. His clear awareness of thermal, solar, and physical principles and his experienced ingenuity have resulted in an integrated house that includes an elegant but simple airflow system and, as William Shurcliff says, a "galaxy" of other innovations.*

Excerpts from "Saunders' Shrewsbury House" by William A. Shurcliff:

Shrewsbury House is a large (2450sf) two-story house in a cold part of Massachusetts, 100% solar heated and built at a cost not significantly different from that of a conventional house of similar size.

All of the rooms, including north rooms, remain at about 70F throughout the winter, even in week-long cloudy periods. There are generously large window areas on all four sides of the house. There is a steady flow of fresh air into the house 24 hours a day.

Persons in the living area may be entirely unaware that the house is solar heated. Operation in winter is fully automatic. As far as temperature control is concerned, the occupants have no duties. No thermal shades to operate — there are no thermal shades. Humidity is well controlled.

The large integral greenhouse never cools down to 32F, yet it employs no auxiliary heat and has no added thermal mass — and it is single glazed.

In summer the house keeps exceptionally cool, even in long sunny hot spells.

## My First View of the House

On March 3, 1982 I was privileged to be given a slow, thorough, guided tour of the Shrewsbury House by its solar designer Norman B Saunders. I spent two hours inspecting the various components of the house and the solar heating system, and I spent additional hours asking follow-up questions, studying the blueprints of the house, and trying to understand in depth why each component was designed in the given way and how the components work together to provide such an ideal performance.

## Shrewsbury House
Massachusetts, 1982
1890 sf inner floor area

When we arrived at the house, it was empty. Nobody home. Outdoor temperature: 20F. Indoor temperature: 70F. A multi-point recorder showed that all main parts of the living area were close to 70F.

The attic was unbearably hot: 100F. This was, of course, good: a half-million Btu of heat was in storage there, as explained later.

All rooms were brilliantly daylighted, but without glare.

From every room I could enjoy a view in at least two directions.

The greenhouse was at about 70F.

I walked into every room — and saw no sign of any solar heating system.

The attic, however, was bristling with solar equipment.

To devote an entire book to one solar-heated house may seem absurd, but the Shrewbury House is very different, a wholly new approach to solar heating. A galaxy of new concepts is involved.

The strategies that Saunders used in planning the solar heating system for the Shrewsbury House are, first:

> Variety: Use a great variety of components and processes that are carefully matched to one another so as to work together cooperatively, each making up for the limitations of others. Each may be simple, inexpensive, and not especially impressive; but together, they provide superb performance.

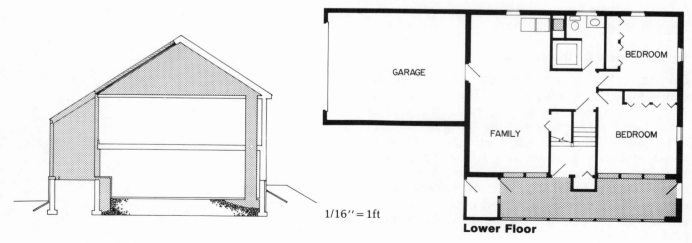

1/16″ = 1ft

**Lower Floor**

True integration and double-duty function: An attic south roof, if properly designed, can serve not only to shed rain but also to admit solar radiation. Attic insulation can serve not only to insulate the home in general but also to insulate an upper thermal storage system. Insulated foundation walls can serve not only to support the house but also to contain and insulate a lower thermal storage system. Vertical south windows, if properly designed, can serve not only for admitting light and view but also for air-thermo-syphoning collection of solar energy. A single fan can serve not only to drive hot air from an upper storage system to a lower storage system but also to supply warm air to a greenhouse and drive stale air to the outdoors via an air-to-air heat exchanger and induce a corresponding inflow of fresh air. Making major components of the house do double duty saves money, saves space, and minimizes complexity.

Other general strategies include:

Taking the greatest pains to reduce heat loss.

Providing two large-thermal-capacity storage systems: one (using water) above the upper story at a temperature well above room temperature, and the other (using stone) below the lower story at about 70F.

Maintaining a continuous flow of air from the upper storage system to the lower.

Protecting the large south window area of the house proper with a thermally buffering greenhouse.

### Shrewsbury House

| | |
|---|---|
| 6500 | heating degree-days |
| 50 | percent of possible sunshine |
| $90,000 | cost |
| $48 | square foot cost (inner house only) |
| $42 | square foot cost (incl. greenhouse) |
| 1890 | sf inner house floor area |
| 259 | sf greenhouse area |
| 235 | sf vertical south glass area |
| 235+410 | sf sloping south glass area |
| 0.46 | ratio of south glass to inner floor |
| 84 | sf non-buffered glass area |
| 10 | sf buffered glass not on the south |
| | ft envelope height |
| R-11 | ceiling insulation |
| | first floor insulation |
| | inner of double walls insulation |
| R-30 | single wall insulation |
| | outer of double walls insulation |
| R-45 | roof insulation |
| | foundation insulation |
| high | estimated overall tightness |
| | Btu/DD/sf |
| | Million Btu's fuel consumed |
| none | type of auxiliary heat |
| | estimated comfort level |
| | maximum winter greenhouse temp. |
| | minimum winter greenhouse temp. |

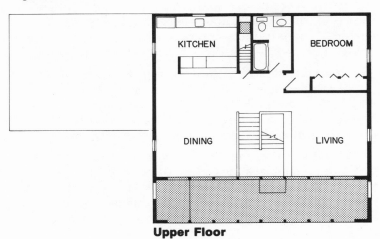

**Upper Floor**

Arranging these south window assemblies as thermosyphoning collectors, delivering warm air to the attic.

Arranging for the system to be self-regulating.

The specific implementing strategies include:

Upper solar-energy-collecting-and-storing (attic) system.

Lower thermal storage system (bin of stones).

Greenhouse, single glazed.

Air-drive system, using a single 1/4hp fan but many flows.

The most important (and most remarkable) component of the Shrewsbury House solar heating system is the attic solar window. Because it employs a set of louvers that somewhat correspond (in size, spacing, and slope) to the treads of a staircase, this window has been called by its inventor a Solar Staircase™. The name is certainly apt, and it has been trademarked. The window system as a whole has been patented.

**Typical sunny daytime in winter**

During a typical sunny daytime in winter, solar energy pours into the attic via the attic solar aperture and warms the air there and the water-filled carboys there. Also it pours into the greenhouse, warming the greenhouse earth and the greenhouse air — and some pours into the upward projecting south foundation wall of the greenhouse, warming this wall. Some pours through the big south windows of the south rooms, warming these rooms by direct radiation, and some is absorbed within those windows and warms the air there and causes this to rise, by thermosyphon, into the attic.

The fan runs steadily, transporting hot air from the upper storage system to the lower storage system and forcing an airflow from the latter storage system to the greenhouse and thence via a special duct (part of a heat exchanger) to the outdoors.

At the same time, the slight negative pressurization of the house causes an inflow of cold outdoor air via the heat exchanger, and this (still fairly cold) air enters the upper southeast room, diffuses to the other rooms of both stories, and eventually finds its way into the attic.

All vents and all exterior doors and windows are closed. The fan automatically speeds up, speeding delivery of fresh cool air to the rooms and delivery of slightly warm air to the greenhouse, whenever (1) the rooms tend to become too hot or (2) the greenhouse threatens to cool down to 32F.

All of the rooms are at about 70F, and the humidity is in an acceptable range, not far from 40%RH. There is a steady inflow of fresh air. All rooms have good daylighting, without glare. Everything is silent; there are no moving parts other than the fan (audible only if you are standing within a few feet of it), and air.

### Typical cold night in winter

The fan continues to run, transporting warm air from upper storage system to lower storage system, and driving cool air (at about 50F, say) from there to the greenhouse, and driving greenhouse air to the outdoors, and causing a steady inflow of cold fresh air to be delivered to the upper southeast room.

All vents (including the greenhouse vents) and all exterior doors and windows remain closed.

Operation is fully automatic. The fan continues to run at whatever speed the control system judges to be appropriate. There are no thermal shades or shutters for the house occupants to close.

All of the rooms remain at about 70F, and the humidity remains satisfactory.

The greenhouse cools down only very slowly, thanks to the large amounts of heat stored in the greenhouse earth and the greenhouse south foundation wall, and thanks also to the stream of slightly warm air (at about 50F) flowing steadily from the lower storage system to the greenhouse. It is helpful also that the coldest air in the greenhouse — the air descending along the single south glazing — is preferentially collected and ejected.

Of course, the overall excellent insulation of the house helps retain heat, and the greenhouse acts as a buffer region for the big south windows of the living area. The overall airtightness helps also.

## Nyholm House
Nebraska, 1980
1536 sf inner floor area

*This three-story envelope house has vertical glass on its entire south face, giving it a south glass/interior floor area ratio of .56, which is among the highest of all the houses in our survey. With such a large glass area we would normally expect the temperature swings in the greenhouse to be excessive, but this has apparently not happened — the owner has reported a greenhouse winter minimum of 36F and a winter maximum of 76F.*

Comments by Paul Nyholm, designer and owner:

I just wanted a maintenance free house and low cost utilities, so the entire house is cedar, inside and out, with a natural finish. It is surrounded by creeping red fescue grass, which does very well in the shade. It is never cut, but still offers a natural good look. Aesthetically I think the house fits in well, windows reaching for the sun and arched by trees.

The site is very steep and wooded; it drops thirty-five feet into a creek. I took soil tests when we started, but the footings ran into fill dirt five feet from a test hole so had to move the house twenty-four feet at the last moment. I used avany concrete block with steel (horizontal and vertical) and then slushed, with foam outside the block.

Vertical glass covers a twenty-four by forty foot southern exposure. I didn't need that much glass to keep the house efficient, but I wanted the expanse to fit my taste. A small natural gas furnace serves as backup. Except when I turn it on for about ten minutes in the morning, the furnace has never

1/16'' = 1ft

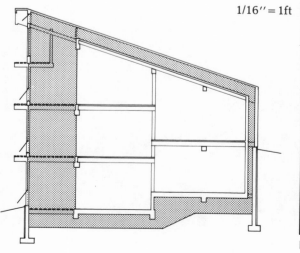

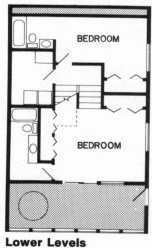

**Lower Levels**

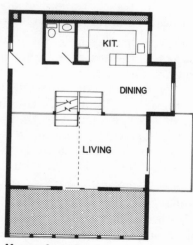

**Upper Levels**

operated in the daytime, and has run only a little at night. My total heat bill for last year was $119.00.

A few problems have occurred with humidity in winter as well as summer. I have a hot tub in the greenhouse that is kept at 100F. The outside windows were frosting at night and then melting and running down in the daytime. I put a Styrofoam blanket and a sheet of plastic on the hot tub. Then I covered the ground in the crawlspace with plastic, and the problem was solved.

Last summer I traveled a great deal, and was in and out of the house. I left the high greenhouse windows and most of the windows into the house open all summer. This year I will only exhaust the entire house for four to six weeks in the spring and four to six weeks in the fall (when the humidity is at its lowest). Otherwise I will keep the house pretty much shut up and use the Casablanca fan.

I have three eight-inch pipes underground that extend from the north wall to the south wall and plan to put a small fan in each one. Thereby I hope to draw some cool air from the ground. It is very humid in Omaha. I am single so no one is at home during the day, and the house would operate better if someone were there to open the atrium's inner doors to let heat in the house in winter as well as to vent it in the summer.

I am very happy with the house (five levels in all), the way one space leads to another, the vertical heights in rooms. It is a great place to entertain. Most of all, however, it is part of me.

Looking southwest from kitchen.

### Nyholm House

| | |
|---:|:---|
| 6100 | heating degree-days |
| 60 | percent of possible sunshine |
| $86,000 | cost |
| $55 | square foot cost (inner house only) |
| $50 | square foot cost (incl. greenhouse) |
| 1536 | sf inner house floor area |
| 192 | sf greenhouse area |
| 864 | sf vertical south glass area |
| 0 | sf sloping south glass area |
| 0.56 | ratio of south glass to inner floor |
| 0 | sf non-buffered glass area |
| 0 | sf buffered glass not on the south |
| 24 | ft envelope height |
| R-20 | ceiling insulation |
| R-20 | first floor insulation |
| R-11 | inner of double walls insulation |
| R-19 | single wall insulation |
| R-11 | outer of double walls insulation |
| R-20 | roof insulation |
| R-10 | foundation insulation |
| | estimated overall tightness |
| 0.75 | Btu/DD/sf |
| 7.0 | Million Btu's fuel consumed |
| gas | type of auxiliary heat |
| normal | estimated comfort level |
| 75 | maximum winter greenhouse temp. |
| 36 | minimum winter greenhouse temp. |

## Demmel House
Nebraska, 1980
1600 sf inner floor area

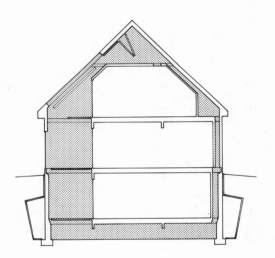

*This is an envelope house with insulating shutters on the sloping greenhouse glass and an earth tube for summer cooling. Its performance has been monitored by Bing Chen and others at the University of Nebraska, and some of the results are shown in a Chapter 3 chart. Heat stratification in the envelope spaces was a problem until a fan was installed; summer cooling was also a challenge. Heating costs for this house have been outstandingly low.*

Comments by Dennis Demmel, owner:

The home of Dennis and Ruth Demmel was patterned after the Mastin house, a thermal envelope design in Rhode Island. This three-level home has a total of 1600sf of finished living area. The main (middle) level is at grade, and the soil level of the crawlspace is nearly 11ft below grade.

Insulation above grade is fiberglas, with 6″ in outer walls and roof, 3-1/2″ in inner walls and 6″ in the ceiling. Insulation below grade is a combination of Styrofoam and Thermax insulation for an R-value of 16. This extends 10ft-8″ down on the east and west walls. On the north and south, the concrete block walls are insulated to a 4ft depth and then horizontally 4ft out, again with R-16. Vapor barriers were used on the inside of both inner and outer shells of the house.

The south glazing is nearly 300sf in the roof above the solarium and 150sf in the solarium south wall. Only one small window was installed on the north of the house, primarily for ventilation.

1/16″ = 1ft

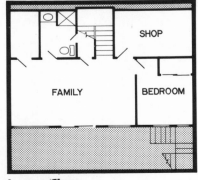

**Lower Floor**

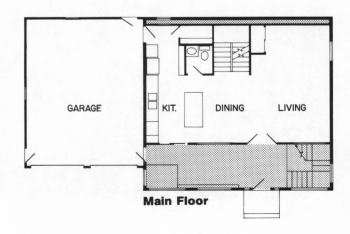

**Main Floor**

Since attic temperatures exceeded 100F on sunny winter days and created a major temperature difference between the attic and the outdoors, a fan was utilized to move heat down from the attic through all three levels of the living area and eventually into the crawlspace. Considerably greater heat transfer into the soil was indicated, particularly at 6″ depth, when compared to the transfer with natural convection only.

The use of the fan to move attic heat has several advantages in addition to improving total system efficiency. The solarium is kept much cooler on sunny days, which can be valuable for plant growth conditions and for human habitation. In addition, the fan can provide solar heat to areas of the interior that would otherwise not be heated. Windows and doors between solarium and living area can also provide some of that heat transfer when the occupants are home, but in many homes (such as our own) the family members often leave in the morning when it is too cool to open window or doors to the solarium, and may return later in the day after the optimum time for closing them. This creates significant additional heat loss to the solarium, and an automated fan and shutter system are obviously beneficial.

Temperatures on the lower level during the first winter were often lower than desired. Part of the reason for this is the open stairwell which permits heat stratification, giving a temperature difference as much as 15F between upper and lower levels of the living space. One solution considered is the installation of a low-cfm fan to move warmer air from the ceiling of the upper level to the floor of

## Demmel House

| | |
|---|---|
| 7200 | heating degree-days |
| 55 | percent of possible sunshine |
| $70,000 | cost |
| $43 | square foot cost (inner house only) |
| $38 | square foot cost (incl. greenhouse) |
| 1600 | sf inner house floor area |
| 256 | sf greenhouse area |
| 150 | sf vertical south glass area |
| 300 | sf sloping south glass area |
| 0.28 | ratio of south glass to inner floor |
| 68 | sf non-buffered glass area |
| 15 | sf buffered glass not on the south |
| 34 | ft envelope height |
| R-20 | ceiling insulation |
| R-13 | first floor insulation |
| R-12 | inner of double walls insulation |
| R-20 | single wall insulation |
| R-20 | outer of double walls insulation |
| R-20 | roof insulation |
| R-16 | foundation insulation |
| | estimated overall tightness |
| 0.15 | Btu/DD/sf |
| 1.7 | Million Btu's fuel consumed |
| elec | type of auxiliary heat |
| normal | estimated comfort level |
| 100 | maximum winter greenhouse temp. |
| 33 | minimum winter greenhouse temp. |

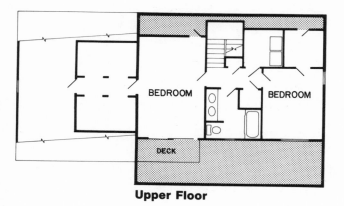

**Upper Floor**

The Demmel house is similar to the Mastin House in Rhode Island, shown here.

the lower level whenever the temperature difference exceeds about 5F degrees. This approach would also lower heat loss at the upper level ceiling by lowering the temperature there. Other considerations regarding heat stratification include moving heat from the clothes dryer and the refrigerator both down one level. In future designs it is suggested that the envelope home be restricted to two levels rather than three, and that the open stairwell should be avoided.

This home has a number of features utilized in the cooling mode during the summer months. Gable end vents are located at the east and west ends of the attic and are 5ft high by 10ft wide triangular vents. These vents are open most of the summer. The upper level living area has ceiling vents to vent warm ceiling air into the attic at night. Small north doors allow air from between the double north walls to enter and assist in cooling during summer nights. Since the upper level has no north or west windows for cross-ventilation, the ceiling vents provide a good alternative, along with the north doors. Ventilation has been excellent.

A cooling tube located under the back lawn is a culvert 80ft long and 2ft diameter made of corrugated aluminum. The upright portion at the far end is screened and allows outside ventilation air into the tube for cooling and replacing heated air drawn from the top of the house. The inlet is located to the northeast of the house and has a coned top with shroud around the perimeter to minimize the venturi effect of cross winds. Some experimentation was conducted using a fan to move cooling tube air through the home during hot summer evenings. This seems to be more effective than the natural convection. Since this house has no forced air system, there can be times of some discomfort which air circulation would help to minimize, and a fan would be helpful just to circulate air for evaporative cooling.

With the elimination of the roof glass above the solarium and the use of proper eave lengths above each row of vertical glazing on the south wall, it is probable that the summer cooling load would be reduced greatly. Therefore ventilation requirements would be reduced and could be nearly eliminated during the day. The cooling tube should be used only when necessary at night, to conserve its cooling capacity for a full summer. A closed loop cooling tube is also recommended, particularly if a fan is used anyway.

Although various difficulties have been encountered with the thermal envelope design, this author believes that the system has tremendous potential with appropriate changes made. The great amount of mass under and around the base of the house is nearly free, and has shown considerable response to temperatures.

Last January we had almost no sun, and very cold record-breaking weather. As a result, we were using more electric heat. However, $20 was the actual heating cost for the whole winter.

## Stewart House
California, 1980
960 sf inner floor area

*The north wall airspace in this envelope house was made three feet wide in order to make it more useful: for closets, a stairway, and storage. This plan departs from the common practice of keeping flammable materials out of the airspaces as much as possible.*

Comments by Bob Hoagland, designer and builder:

To save space and the cost of another wall the whole north plenum was made 3 feet wide. Downstairs it was bedroom closets, entry hall and staircase. Upstairs it was the upper landing of stairs, living room closet and kitchen pantry.

12-inch openings in the floor both north and south were covered with appropriate grates and were the only restriction in the plenum. The 13 foot staircase has 2x10 treads and no risers. At the head of the stairs there was placed a 34″ by 76″ dual-pane fixed window. The purpose of placing the window there was, in addition to lighting the stairwell, to further cool the air coming down the north plenum, making it more dense and aiding the flow around the loop.

A heating stove was never installed. At about 3000 degree-days per year, $30/month was the total electric bill for hot water, cooking and heating (at $.05/kWhr).

Because of the size of the lot and the house, it was impossible to face the greenhouse glass due south. The SSE exposure to the summer sun makes shading with an overhang practically impossible, and it therefore became a choice of either overheating or loss of view. Needless to say I shall not design that problem into a house again.

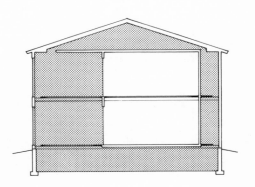

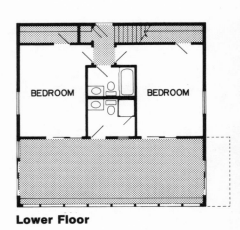

**Lower Floor**

North wall stairway and closets provide airflow passages.

**Stewart House**

| | |
|---:|:---|
| 3000 | heating degree-days |
| 50 | percent of possible sunshine |
| $30,000 | cost |
| $31 | square foot cost (inner house only) |
| $21 | square foot cost (incl. greenhouse) |
| 960 | sf inner house floor area |
| 448 | sf greenhouse area |
| 324 | sf vertical south glass area |
| 0 | sf sloping south glass area |
| 0.34 | ratio of south glass to inner floor |
| 174 | sf non-buffered glass area |
| 0 | sf buffered glass not on the south |
| 18 | ft envelope height |
| R-19 | ceiling insulation |
| R-19 | first floor insulation |
| R-11 | inner of double walls insulation |
| R-19 | single wall insulation |
| R-19 | outer of double walls insulation |
| R-30 | roof insulation |
| R-11 | foundation insulation |
| | estimated overall tightness |
| 0.42 | Btu/DD/sf |
| 1.2 | Million Btu's fuel consumed |
| elec | type of auxiliary heat |
| normal | estimated comfort level |
| 82 | maximum winter greenhouse temp. |
| 52 | minimum winter greenhouse temp. |

1/16″ = 1ft

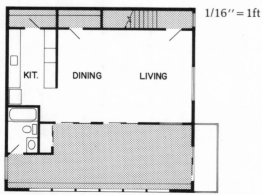

KIT.   DINING   LIVING

**Upper Floor**

## Anawalt House
California, 1979
2400 sf inner floor area

*In this envelope house, the floors and inner walls were left uninsulated, and the greenhouse glazing is only single-layer (double-glazing is most frequently used). It is in a fairly warm (3000 degree-day) California climate.*

*Loop airflows can be dampered to reduce night-time heat loss.*

Comments by Douglas Anawalt, designer:

I experimented with a couple of variations on the convective loop system. One experiment was blocking or dampering the collective leg. For budget reasons, I wanted to single-glaze the solar wall. I also wanted to reduce the loss of stored heat to the cold solar wall at night. The blocked loop seems to noticeably buffer the living spaces and the thermal mass from this heat loss, but not without some condensation on the inside of the solar wall.

I also experimented with the thermal mass. Lee Butler talked me out of using rock-bed storage but I wanted to make an effort to capture some of the heat from the warmer air that moves along the underside of the floor in addition to any "heat sink" effect that may result from the exposed earth in the crawlspace. My solution was to place rock in wire cribbing in a triangular pattern that reached to the underside of the floor joist. I am still undecided on the question of rock or no rock.

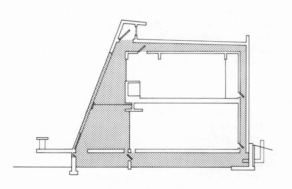

1/16″ = 1ft

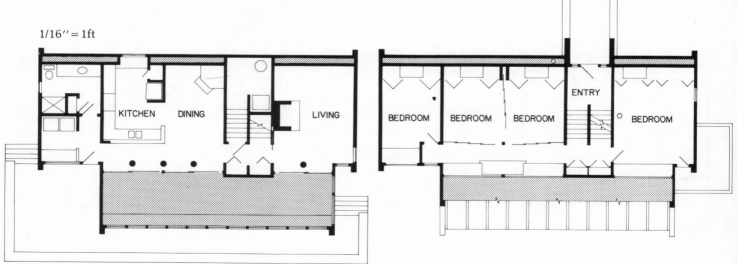

**Lower Floor**

**Upper Floor**

As a contractor, I am particularly interested in framing techniques that reduce the cost of the double shell. This house is located on a site that slopes to the south. I keep my frame independent of the retaining wall at the north elevation. This allows me to balloon frame the exterior wall and make maximum use of the plywood platform. I use wall jacks and lift the exterior walls with windows and siding in place to reduce framing labor. This house has a stucco exterior but was detailed for a tilt-up wood exterior as an alternate.

The interior walls are platform framed with the inner north wall duct lining in place when they are tilted up. R-19 exterior walls have studs 24″ on center to reduce the weight.

One other comment occurs to me concerning the high solar wall. I kept the solar wall glazing high for two reasons:

1. To provide a view through the greenhouse from the second floor rooms.

2. To create a sufficient heat differential during the cooling cycle when the lower portion of the greenhouse is screened to protect the plants.

This is a good example of the need to balance solar requirements with other considerations.

### Anawalt House

| | |
|---|---|
| **3000** | heating degree-days |
| **50** | percent of possible sunshine |
| **$98,000** | cost |
| **$40** | square foot cost (inner house only) |
| **$35** | square foot cost (incl. greenhouse) |
| **2400** | sf inner house floor area |
| **418** | sf greenhouse area |
| **0** | sf vertical south glass area |
| **440** | sf sloping south glass area |
| **0.18** | ratio of south glass to inner floor |
| **30** | sf non-buffered glass area |
| **38** | sf buffered glass not on the south |
| **21** | ft envelope height |
| **R-13** | ceiling insulation |
| **R-0** | first floor insulation |
| **R-0** | inner of double walls insulation |
| **R-19** | single wall insulation |
| **R-19** | outer of double walls insulation |
| **R-19** | roof insulation |
| **R-0** | foundation insulation |
| | estimated overall tightness |
| **2.50** | Btu/DD/sf |
| **18.0** | Million Btu's fuel consumed |
| **wood** | type of auxiliary heat |
| | estimated comfort level |
| **120** | maximum winter greenhouse temp. |
| **50** | minimum winter greenhouse temp. |

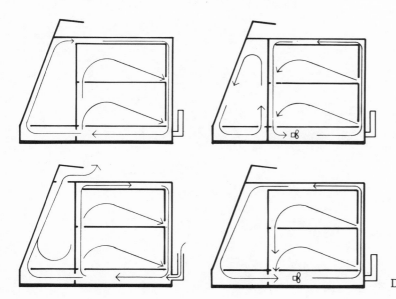

Dampers provide many airflow options.

Identified envelope houses at the time of our survey.

# 9

## SURVEY OF ENVELOPE HOUSES

### How We Obtained the Data

In preparing for writing this book, we obtained data on more than one hundred envelope houses that had been occupied through at least one winter, to find out just how well these controversial buildings were actually performing under real conditions. We sent letters and questionnaires to state energy offices, to owners, designers, and others we knew, to locate as many envelope homes as we could and to obtain key information about the houses' design and performance. We also did a lot of telephoning. After we assembled the preliminary data, we sent a printout of the information back to owners and designers for them to check.

### Remarkable Performance

It is readily apparent from the results of this survey that even the average performance of these houses is remarkable, though many were built by inexperienced builders and designers had few guidelines to follow. The survey represents a wide variety of impressively performing designs in a variety of climates.

There is much more we would like to know, but the picture we did obtain is importantly useful. It makes possible a cumulative assessment of the existing envelope buildings, it provides a background for the description and monitoring and understanding of individual buildings, and its overview prepares for future studies.

### Many Variables

One of the things that we had hoped for in doing this survey was to find correlations between some of the elements covered. For example, what was the correlation between the amount of insulation and the amount of auxiliary heat used, in Btu/DD/sf? Or the correlation between the area of south glass and the greenhouse minimum temperature?

It quickly became obvious that the number of variables involved was too large for most of those simple correlations to show up. The houses differed so much and in so many ways that to look at two or three elements by themselves was to ignore many other relevant factors, some of which were in our data and some of which were not. A surprisingly large proportion of houses did not conform to any conventional single definition of what an envelope house is.

Some of the information we obtained from owners is not precise: temperatures were taken at different locations and times, quantities of fuel for auxiliary heating were determined in many different ways, the effectiveness of infiltration control could not be measured, and the living habits of occupants vary in unknown ways.

### Full Survey Chart

The chart on the following pages presents nearly all of the data we obtained for 108 houses. At the bottom of the first pages there is an explanation of each of the columns and how the information for each was derived.

The top two lines on each pair of data pages show the mean and median values for each of the columns. The mean is the conventional average, obtained by adding all the values in the column and then dividing by the number of houses for which we had that data. The median is the "middle" value of each category; half the values are less than or equal to the median and half are greater than or equal to.

| Location | Fahrenheit Degree-Days | Percent Sunshine | Owner Cost in Thousands of Dollars | Cost/Square Foot | Inner Floor Area | Greenhouse Area | Vertical South Glass | Sloped South Glass | Glass/Floor Ratio | Direct Glass | Other Buffered Glass | Buffer Height | R-Values of Insulations | | | | | | |
|---|---|---|---|---|---|---|---|---|---|---|---|---|---|---|---|---|---|---|---|
| | | | | | | | | | | | | | Ceiling | Floor | Inner Walls | Single Walls | Outer Walls | Roof | Foundation |
| MED.: | 6700 | 50 | 75 | 43 | 1800 | 300 | 228 | 170 | 0.23 | 48 | 24 | | 19 | 13 | 11 | 20 | 19 | 20 | 11 |
| MEAN: | 6400 | 49 | 80 | 44 | 1820 | 335 | 252 | 194 | 0.25 | 56 | 28 | | 17 | 13 | 12 | 22 | 20 | 24 | 12 |
| MAA | 6000 | 45 | 130 | 51 | 2514 | 308 | 239 | 226 | 0.22 | 75 | 58 | 34 | 19 | 19 | 11 | 32 | 22 | 19 | 16 |
| MAB | 6000 | 45 | 64 | 53 | 1200 | 200 | 96 | 72 | 0.14 | 48 | 6 | 27 | 19 | 19 | 11 | 22 | 19 | 30 | 5 |
| MAC | 5600 | 55 | 53 | 26 | 1972 | 0 | 120 | 0 | 0.06 | | 101 | 31 | 20 | 20 | 13 | 20 | 13 | 20 | 10 |
| MAD | 5600 | 55 | 30 | 20 | 1440 | 270 | 200 | 192 | 0.27 | 38 | 7 | 30 | | | | | | | |
| MAE | 6500 | 55 | 122 | 54 | 2240 | 284 | 144 | 356 | 0.22 | 200 | 30 | 27 | 16 | 11 | 11 | 30 | 30 | 19 | 10 |

## EXPLANATION OF COLUMN HEADINGS ON FULL SURVEY CHART

### Location

First two letters are abbreviations for state where house is located. 'CD' is used for all Canadian houses.

### Fahrenheit Degree-Days

Heating degree-days are derived by totaling, for each day of the year, the difference between the average temperature for the day and 65F (if the average is below 65F). This is based on the assumption that auxiliary heat will be required to maintain comfort levels whenever the outdoor temperature is below 65F, an assumption that becomes less and less valid as a structure is better insulated. For example, a super-insulated house would usually not require any auxiliary heating when outside temperatures are down to 40-50F.

Degree-day values used here are 30-year averages from the Environmental Science Service Administration (ESSA) weather maps, except where more accurate values for the winter of 1980-81 were available.

### Percent Sunshine

This is the mean percent of possible sunshine for the winter months of December through February, interpolated from ESSA maps. The percent sun refers to the portion of the day when the sun is shining, not obscured by clouds. Percent sun is not directly proportional to the percentage of solar gain available, since for example on a day with zero percent sun (completely cloudy) there can still be 20 to 30 percent of the possible solar radiation available.

### Owner Cost in Thousands of Dollars

Cost to owners in 1980 dollars, as reported by survey respondents for unfurnished houses excluding design, appliances, and site costs of land, water supply, sewage disposal, driveway, and landscaping. No attempt was made to adjust values for owners' labors.

### Cost/Square Foot

Owner cost divided by gross inner house floor area (see next item).

### Inner Floor Area

Gross floor area of inner house, excluding sunspace/greenhouse and any other buffering airspaces. Basements were also not included with the inner floor area.

### Greenhouse Area

Gross area based on width and depth, excluding additional decks.

### Vertical South Glass

Square feet of vertical south aperture.

### Sloped South Glass

Square feet of sloping south aperture.

### Glass/Floor Ratio

Total south aperture of greenhouse/sunspace divided by the floor area of the inner house.

### Direct Glass

Non-buffered glass areas, directly between inner house and outdoors. Usually on the east and west.

| Location | Tightness | Btu/DD/sf | Millions of Btu | Type of Heat | Comfort Level | Max. GH Temperature | Min. GH Temperature | |
|---|---|---|---|---|---|---|---|---|
| | | 0.96 | 12.00 | | | 90.0 | 40.0 | |
| | | 1.29 | 14.84 | | | 91.5 | 41.2 | |
| MAA | | 0.40 | 6.00 | W | C | 85.0 | 33.0 | FANS, JAPANESE GARDEN GREENHOUSE ('TORII') |
| MAB | M | 4.44 | 32.00 | W | N | 98.0 | 42.0 | RETROFIT, WARM WITH INCOMPLETE INSULATION |
| MAC | | 1.09 | 12.00 | W | | | 45.0 | ENVELOPE AT EAST, WEST, SOUTH WINDOWS ('FAWCETT') |
| MAD | M | 0.83 | 6.70 | | | 80.0 | 37.0 | UNCOMPLETED. SPRINKLER FIRE PROTECTION |
| MAE | | 0.62 | 9.00 | W | N | 95.0 | 40.0 | CHEVRON SHAPE, SOME DIRECT GAIN ('MINERGY') |

## Other Buffered Glass

Buffered glass areas other than on the south: double windows with a buffering airspace in between. Usually on the north, in double north walls of envelope houses, but some houses had buffered windows on the east and west.

## Buffer Height

Interior overall height of all the buffer spaces. This is the difference in height between the bottom of the lowest buffer space and the top of the highest.

## R-Values of Insulations at Specified Locations:

**Ceiling** — top of inner shell, under attic buffer airspace.

**Floor** — bottom of inner shell, over basement or crawlspace.

**Inner Walls** — walls between inner house and buffering airspaces.

**Single Walls** — walls directly between inner house and outdoors.

**Outer Walls** — walls between buffer airspace and outdoors.

**Roof** — between the upper airspace and the roof.

**Foundation** — usually rigid polystyrene on exterior.

## Tightness

Rough evaluation of the tightness of the house, based on respondents' reports of quality of air/vapor barrier installation, fit of windows, draftiness, stuffiness, etc.

## Btu/DD/sf

Total auxiliary heat expressed in Btu (see next category), divided by the Fahrenheit degree-days for the location and by the floor area of the inner house (the heated space only, as above). The purpose of this calculation is to provide a basis for comparing the heating needs of different homes apart from their differences in size and climate, but the method has deficiencies which are discussed in the pages following the survey chart.

## Millions of Btu

Estimated Btu value of the auxiliary energy used, based on survey respondents' reports. This only includes energy used specifically for heating, and does not include heat from appliances and other internal gains. For wood, we assumed a yield of 12,000,000 Btu per cord of hardwood and 8,000,000 Btu per cord of soft wood, burned in a moderately airtight stove. Where electric heat was used, figures are shown only for houses in which we could separate the amount used for heating from the total amount of electricity used.

## Type of Heat

Primary auxiliary heat source:

W = wood
E = electricity
O = other

## Comfort Level

The comfort levels at which auxiliary heat is maintained.

W = warm. Thermostat or stove maintains steady warm temperatures.

N = normal. The usual management of allowing the house to cool down at night and while owners are away at work during the day.

C = cool. Little or no auxiliary used. Temperature swings are accepted and lived with, for example by adjusting clothing.

## Maximum Greenhouse Temperature

Maximum winter temperature recorded in the greenhouse or sunspace, avoiding direct sunshine. Locations were not consistent.

## Minimum Greenhouse Temperature

Minimum temperature recorded in the greenhouse or sunspace. Locations were not consistent.

| Location | Fahrenheit Degree-Days | Percent Sunshine | Owner Cost in Thousands of Dollars | Cost/Square Foot | Inner Floor Area | Greenhouse Area | Vertical South Glass | Sloped South Glass | Glass/Floor Ratio | Direct Glass | Other Buffered Glass | Buffer Height | R-Values of Insulations | | | | | | |
|---|---|---|---|---|---|---|---|---|---|---|---|---|---|---|---|---|---|---|---|
| | | | | | | | | | | | | | Ceiling | Floor | Inner Walls | Single Walls | Outer Walls | Roof | Foundation |
| MED.: | 6700 | 50 | 75 | 43 | 1800 | 300 | 228 | 170 | 0.23 | 48 | 24 | | 19 | 13 | 11 | 20 | 19 | 20 | 11 |
| MEAN: | 6400 | 49 | 80 | 44 | 1820 | 335 | 252 | 194 | 0.25 | 56 | 28 | | 17 | 13 | 12 | 22 | 20 | 24 | 12 |
| RIA | 5800 | 50 | 86 | 45 | 1884 | 264 | 144 | 376 | 0.28 | 24 | 24 | 30 | 11 | 11 | 11 | 19 | 19 | 19 | 12 |
| RIB | 5800 | 50 | 57 | 47 | 1188 | 224 | 210 | 16 | 0.20 | 50 | 52 | 30 | 19 | 19 | 11 | 19 | 19 | 19 | 10 |
| NHA | 7500 | 50 | 80 | 69 | 1150 | 396 | 300 | 0 | 0.26 | 9 | 35 | 19 | 19 | 19 | 19 | 19 | 19 | 30 | 19 |
| NHB | 7300 | 50 | 60 | 36 | 1632 | 256 | 180 | | 0.11 | 41 | 17 | 30 | 19 | | 11 | 24 | 24 | 19 | 15 |
| NHC | 7300 | 50 | 120 | 55 | 2154 | 432 | 0 | 504 | 0.23 | 85 | 40 | 20 | 13 | 8 | 13 | 19 | 26 | 35 | 13 |
| NHD | 7500 | 50 | 90 | 34 | 2600 | 256 | 332 | 126 | 0.18 | 40 | 0 | 26 | 19 | 19 | 13 | 19 | 19 | 30 | 10 |
| NHE | 7500 | 50 | | | 1440 | 278 | 340 | 32 | 0.26 | 64 | | 26 | 22 | 11 | 11 | 22 | 19 | 22 | 10 |
| NHF | 7500 | 50 | 66 | 34 | 1924 | 237 | 250 | 0 | 0.13 | 48 | 52 | 29 | 19 | 11 | 11 | 22 | 22 | 22 | 15 |
| NHG | 7700 | 50 | 40 | 20 | 2000 | 400 | 218 | 176 | 0.20 | 0 | 27 | 23 | 11 | 19 | 11 | 11 | 11 | 19 | 10 |
| NHH | 7500 | 50 | | | 1344 | 224 | 164 | 90 | 0.19 | | | 28 | 13 | 13 | 13 | 21 | 21 | 21 | 10 |
| NHI | 7600 | 45 | 69 | 71 | 960 | 240 | 366 | 189 | 0.58 | 48 | 18 | 24 | 19 | 19 | 11 | 19 | 19 | 38 | 10 |
| NHJ | 7300 | 50 | 40 | 24 | 1664 | 320 | 156 | 350 | 0.30 | 24 | 24 | 31 | 11 | 11 | 11 | 23 | 23 | 18 | 18 |
| NHK | 7500 | 50 | 35 | 30 | 1152 | 144 | 174 | 116 | 0.25 | 83 | | 22 | 19 | 19 | 13 | 26 | 18 | 19 | 15 |
| NHL | 7300 | 50 | | | 1200 | 180 | | | | | | | | | | | | | |
| NHM | 7300 | 50 | | | 858 | 165 | 384 | 0 | 0.45 | 0 | 32 | 20 | 38 | 19 | 13 | 19 | 19 | | 10 |
| NHN | 7700 | 50 | 70 | 63 | 1100 | 200 | 350 | | 0.32 | 8 | 30 | 23 | 19 | 13 | 13 | 22 | 22 | 19 | 12 |
| NHO | 7500 | 50 | 78 | 39 | 2000 | 228 | 589 | 361 | 0.48 | 0 | 28 | 30 | 19 | 19 | 11 | 19 | 19 | 19 | 9 |
| NHP | 7500 | 50 | 110 | 56 | 1932 | 224 | 230 | 92 | 0.17 | 64 | 24 | 28 | 19 | 19 | 19 | 35 | 30 | 30 | 14 |
| NHQ | 7500 | 50 | 110 | 59 | 1861 | 232 | 225 | 115 | 0.18 | 48 | 72 | 27 | 18 | 19 | 18 | 26 | 26 | 26 | 11 |
| NHR | 7500 | 50 | 180 | 65 | 2768 | 400 | 351 | 115 | 0.17 | 58 | 74 | 32 | 19 | 19 | 19 | 30 | 30 | 30 | 14 |
| NHS | 7500 | 50 | 100 | 57 | 1730 | 360 | 230 | 460 | 0.40 | 70 | 27 | 31 | 13 | 13 | 13 | 20 | 19 | 19 | 13 |
| NHT | 7500 | 50 | 115 | 54 | 2100 | 324 | 207 | 414 | 0.30 | 85 | 33 | 33 | 13 | 13 | 13 | 20 | 19 | 19 | 13 |
| NHU | 7000 | 55 | 105 | 27 | 3801 | 390 | 243 | 146 | 0.12 | 62 | 69 | 26 | 19 | 0 | 13 | 22 | 22 | 48 | 12 |
| MEA | 7800 | 55 | 55 | 36 | 1518 | 408 | 0 | 168 | 0.11 | 62 | 6 | 23 | 19 | 0 | 11 | 24 | 16 | 19 | 5 |
| VTA | 8000 | 40 | 85 | 34 | 2500 | 380 | 111 | 309 | 0.17 | 10 | 30 | 32 | 19 | 11 | 11 | 19 | 19 | 30 | 10 |
| VTB | 8000 | 40 | 70 | 51 | 1372 | 380 | 320 | 0 | 0.23 | 45 | 24 | 26 | 16 | 12 | 16 | 28 | 28 | 28 | 12 |
| VTC | 8000 | 40 | 30 | 20 | 1482 | 480 | 160 | 144 | 0.21 | 34 | 0 | 28 | 19 | 19 | 11 | 19 | 19 | 19 | 0 |
| VTD | 8300 | 35 | 30 | 30 | 1000 | 192 | 97 | 105 | 0.20 | 92 | 0 | 32 | 19 | 19 | 10 | 19 | 19 | 38 | 10 |
| VTE | 7200 | 40 | 55 | 26 | 2080 | 280 | 258 | 214 | 0.24 | 18 | 54 | 36 | 19 | 0 | 11 | 27 | 27 | 27 | 16 |
| CTA | 6400 | 55 | 130 | 59 | 2184 | 286 | 449 | 0 | 0.21 | 0 | 66 | 16 | 11 | 19 | 11 | | 19 | 30 | 7 |
| CTB | 6400 | 55 | 86 | 41 | 2092 | 352 | 540 | 0 | 0.26 | 0 | 72 | 12 | 11 | 19 | 11 | | 19 | 30 | 7 |
| CTC | 6500 | 55 | 140 | 70 | 2000 | 360 | 316 | 0 | 0.16 | 0 | 139 | 12 | 11 | 19 | 11 | | 19 | 30 | 7 |
| CTD | 6400 | 55 | 85 | 42 | 2000 | 260 | 260 | 0 | 0.13 | 0 | 66 | 16 | 11 | 19 | 11 | | 19 | 30 | 7 |
| CTE | 6000 | 55 | 70 | 35 | 2000 | 192 | 342 | | 0.17 | 87 | 0 | 28 | 19 | 0 | 11 | 29 | 19 | 30 | 10 |
| CTF | 5600 | 50 | 19 | 63 | 300 | 100 | 200 | 0 | 0.67 | 0 | 0 | 18 | 16 | 22 | 22 | 22 | 22 | 22 | 12 |

| Location | Tightness | Btu/DD/sf | Millions of Btu | Type of Heat | Comfort Level | Max. GH Temperature | Min. GH Temperature | |
|---|---|---|---|---|---|---|---|---|
| | | 0.96 | 12.00 | | | 90.0 | 40.0 | |
| | | 1.29 | 14.84 | | | 91.5 | 41.2 | |
| RIA | M | 1.46 | 16.00 | E | N | 110.0 | 37.0 | BROOKHAVEN-MONITORED MASTIN HOUSE |
| RIB | M | 2.20 | 15.17 | E | W | 98.0 | 26.0 | LOW-GAIN |
| NHA | | 2.78 | 24.00 | W | N | 80.0 | 42.0 | SINGLE STORY, LONG EAST-WEST |
| NHB | | 2.01 | 24.00 | W | W | 80.0 | 38.0 | WOOD FIRE MAINTAINED MOST OF WINTER |
| NHC | M | 1.14 | 18.00 | W | N | 120.0 | 38.0 | FAN, CONCRETE BLOCK STORAGE, ATTACHED "SILO" |
| NHD | | 0.51 | 10.00 | E | N | 86.0 | 38.0 | LOW IRON TRIPLE-GLAZING ON SOUTH ('MORRISON') |
| NHE | | | | | C | | | |
| NHF | H | 0.62 | 9.00 | W | N | 71.0 | 37.0 | |
| NHG | L | 1.56 | 24.00 | W | W | 95.0 | 25.0 | NOT TIGHTLY CLOSED IN LAST WINTER |
| NHH | | 1.49 | 15.00 | W | N | 85.0 | 32.0 | SIMILAR TO TOM SMITH HOUSE |
| NHI | M | | | W | C | | | |
| NHJ | M | 0.74 | 9.00 | W | N | 80.0 | 45.0 | POST AND BEAM WITH POLYSTYRENE INSULATION |
| NHK | L | 0.35 | 3.00 | W | N | 100.0 | 36.0 | LOW COST 2-STORY ('BEALE') |
| NHL | M | 2.05 | 18.00 | W | | 100.0 | 45.0 | |
| NHM | M | 0.96 | 6.00 | W | N | 140.0 | 49.0 | ENVELOPE IS PART OF LARGER BUILDING |
| NHN | H | 0.28 | 2.40 | | N | 90.0 | 40.0 | EARLY MONITORED 1-STORY ('BURNS') |
| NHO | M | | | E | C | 100.0 | | FAN, NOT OCCUPIED |
| NHP | H | 1.66 | 24.00 | W | W | 90.0 | 38.0 | |
| NHQ | H | 0.64 | 9.00 | W | N | 90.0 | 40.0 | SECOND FLOOR INSULATED FROM FIRST |
| NHR | H | 0.58 | 12.00 | W | N | 90.0 | 40.0 | |
| NHS | M | 1.14 | 18.00 | W | | 118.0 | 38.0 | FAN |
| NHT | M | 1.39 | 18.00 | W | W | 110.0 | 37.0 | FAN, INS UNDER BASEMENT FLOOR, HIGH WATER TABLE |
| NHU | M | 0.45 | 12.00 | | | 80.0 | 51.0 | |
| MEA | | 3.12 | 37.00 | W | | 110.0 | 40.0 | TRIPLE GLAZING EAST AND WEST |
| VTA | L | 1.05 | 21.00 | W | N | 85.0 | 28.0 | SHADED SITE, CHIMNEY ON OUTSIDE WALL |
| VTB | L | 2.19 | 24.00 | W | W | 80.0 | 31.0 | URETHANE INSULATION STRUCTURAL PANELS, UNFINISHED |
| VTC | | 2.36 | 28.00 | W | N | 90.0 | 45.0 | UNINSULATED FLOOR, AIRFLOW OVER GRAVEL FILL |
| VTD | M | 2.89 | 24.00 | W | N | 88.0 | 30.0 | INTERIOR LOOP, ROCK STORAGE INSULATED FROM EARTH |
| VTE | | 1.60 | 24.00 | W | | 90.0 | 35.0 | 2 FT GRAVEL OVER INSULATION ON EARTH |
| CTA | | 0.73 | 10.24 | E | N | 95.0 | 34.0 | 4" AIRSPACE ON ALL 4 SIDES, SHUTTERS ON SOUTH |
| CTB | | 1.05 | 14.00 | E | N | 83.0 | 34.0 | |
| CTC | | | | E | N | 90.0 | 36.0 | FAN, INSULATED SHUTTERS |
| CTD | M | 0.94 | 12.00 | W | N | 65.0 | 30.0 | 4 SIDES BUFFERED ('LAU') |
| CTE | | 3.00 | 36.00 | W | | 80.0 | 50.0 | NO INSULATION IN FLOOR |
| CTF | M | 1.19 | 2.00 | W | | 82.0 | 42.0 | FAN, NIGHT INSULATION, CONCRETE BLOCK STORAGE |

| Location | Fahrenheit Degree-Days | Percent Sunshine | Owner Cost in Thousands of Dollars | Cost/Square Foot | Inner Floor Area | Greenhouse Area | Vertical South Glass | Sloped South Glass | Glass/Floor Ratio | Direct Glass | Other Buffered Glass | Buffer Height | R-Values of Insulations | | | | | | |
| --- | --- | --- | --- | --- | --- | --- | --- | --- | --- | --- | --- | --- | --- | --- | --- | --- | --- | --- | --- |
| | | | | | | | | | | | | | Ceiling | Floor | Inner Walls | Single Walls | Outer Walls | Roof | Foundation |
| MED.: | 6700 | 50 | 75 | 43 | 1800 | 300 | 228 | 170 | 0.23 | 48 | 24 | | 19 | 13 | 11 | 20 | 19 | 20 | 11 |
| MEAN: | 6400 | 49 | 80 | 44 | 1820 | 335 | 252 | 194 | 0.25 | 56 | 28 | | 17 | 13 | 12 | 22 | 20 | 24 | 12 |
| NYA | 5500 | 35 | 185 | 77 | 2380 | 360 | 120 | 280 | 0.17 | 84 | 80 | | 19 | 19 | 11 | 19 | 11 | 19 | 8 |
| NYB | 6700 | 35 | 32 | 25 | 1250 | 128 | 118 | 0 | 0.09 | 26 | 14 | 22 | 38 | 19 | 19 | 23 | 23 | 41 | 18 |
| NYC | 6700 | 40 | 7 | 17 | 408 | 117 | 76 | 36 | 0.27 | 1 | 5 | 22 | 11 | 11 | 11 | 19 | 19 | 19 | 10 |
| PAA | 6000 | 40 | 80 | 30 | 2600 | 300 | 159 | 165 | 0.12 | 32 | 36 | 24 | 14 | 0 | 11 | | 16 | 24 | 14 |
| PAB | 6000 | 35 | 80 | 30 | 2600 | 700 | 470 | 394 | 0.33 | 75 | 162 | | 11 | 16 | 11 | 16 | 19 | 19 | 16 |
| MDA | 4720 | 55 | 100 | 37 | 2652 | 432 | 200 | 468 | 0.25 | 60 | 10 | 40 | 11 | 11 | 11 | 21 | 21 | 30 | 14 |
| MDB | 4600 | 50 | 56 | 28 | 1974 | 268 | 208 | 330 | 0.27 | 25 | 30 | 30 | 30 | 30 | 17 | 25 | 25 | 30 | 13 |
| VAA | 3800 | 50 | 74 | 47 | 1550 | 340 | 450 | 396 | 0.55 | | | 25 | 22 | 19 | 11 | 19 | 11 | 19 | 7 |
| VAB | 3800 | 50 | 80 | 33 | 2410 | 280 | 480 | 504 | 0.41 | 42 | 26 | 22 | 22 | 22 | 13 | 25 | 25 | 25 | 22 |
| WVA | 5500 | 40 | 51 | 44 | 1152 | 167 | 117 | 75 | 0.17 | 12 | 12 | 23 | 19 | 19 | 11 | 19 | 19 | 19 | 11 |
| WVB | 5000 | 40 | 120 | 38 | 3120 | 1200 | 600 | 0 | 0.19 | 160 | 36 | 38 | 38 | 19 | 12 | 26 | 19 | 20 | 19 |
| GAA | 3100 | 50 | | | 1710 | 360 | 347 | 250 | 0.35 | 12 | 48 | | 11 | 11 | 11 | 11 | 11 | 19 | 12 |
| TNA | 3500 | 45 | 42 | 48 | 864 | 288 | 228 | 240 | 0.54 | 30 | 44 | 18 | 19 | 11 | 13 | 20 | 20 | 19 | 11 |
| TNB | 3600 | 40 | | | 1300 | 700 | 0 | 480 | 0.37 | 18 | 40 | 18 | 16 | 13 | 11 | 19 | 19 | 11 | 20 |
| KYA | 4900 | 40 | 110 | 42 | 2600 | | 357 | 200 | 0.21 | 205 | 0 | 20 | 0 | | 0 | 26 | 0 | 30 | 12 |
| OHA | 5500 | 35 | 84 | 44 | 1872 | 336 | 106 | 352 | 0.24 | 75 | 0 | 32 | 27 | 19 | 19 | 32 | 19 | 30 | 13 |
| OHB | 6000 | 35 | | | 2376 | 270 | 75 | 350 | 0.18 | | | 28 | 40 | 19 | 16 | 16 | 16 | 15 | 18 |
| OHC | 6500 | 35 | 65 | 21 | 3008 | 400 | 0 | 540 | 0.18 | 165 | 0 | 31 | 10 | 31 | 11 | 19 | 19 | 31 | 10 |
| MIA | 6900 | 35 | 96 | 53 | 1792 | 256 | 225 | 200 | 0.24 | 96 | 50 | 28 | | 0 | 11 | 21 | 21 | 28 | 10 |
| MIB | 8500 | 30 | 133 | 55 | 2400 | 200 | 281 | 92 | 0.16 | 124 | 9 | 35 | 38 | 0 | 11 | 24 | 24 | 38 | 5 |
| MIC | 8500 | 30 | 60 | 36 | 1664 | 250 | 200 | 30 | 0.14 | 101 | 15 | 29 | 13 | 19 | 13 | 19 | 13 | 30 | 10 |
| IAA | 7000 | 50 | 78 | 45 | 1726 | 256 | 152 | 140 | 0.15 | 16 | 0 | 23 | 19 | 19 | 19 | 24 | 24 | 19 | 10 |
| IAB | 7000 | 50 | 150 | 50 | 2970 | 588 | 335 | 564 | 0.30 | 5 | 40 | 20 | 12 | 6 | 12 | 17 | 17 | 20 | 15 |
| IAC | 6600 | 50 | 67 | 59 | 1120 | 280 | | | | | | | | | 13 | 19 | 19 | | |
| WIA | 7500 | 55 | 125 | 52 | 2384 | 360 | 170 | 270 | 0.18 | 82 | 54 | 30 | 19 | 19 | 11 | 16 | 16 | 30 | 12 |
| MNA | 7800 | 50 | 104 | 52 | 2000 | 450 | 213 | 480 | 0.35 | | | 35 | 60 | 30 | 11 | 11 | 11 | 19 | 5 |
| MNB | 9100 | 45 | | | 1600 | | 408 | 0 | 0.26 | | | | 0 | 0 | 0 | 25 | 28 | 43 | 11 |
| MNC | 10000 | 45 | | | 1728 | 300 | 400 | 260 | 0.38 | 12 | 0 | 42 | 16 | 16 | 11 | 16 | 11 | 30 | 8 |
| SDA | 7800 | 55 | 98 | 43 | 2268 | 368 | 420 | 0 | 0.19 | 13 | 43 | 32 | 19 | 0 | 11 | 22 | 22 | 48 | 11 |
| MTA | 7800 | 30 | 65 | 45 | 1416 | 340 | 323 | | 0.23 | 126 | 24 | 24 | 11 | 2 | 11 | 19 | 11 | 19 | 11 |
| MTB | 7400 | 30 | 117 | 46 | 2494 | 360 | 304 | 384 | 0.28 | 70 | 30 | 33 | 19 | 19 | 11 | 19 | 19 | 19 | 16 |
| ILA | 3800 | 45 | 60 | 42 | 1404 | 450 | 235 | 323 | 0.40 | 39 | 20 | 20 | 11 | 15 | 13 | 19 | 19 | 19 | 11 |
| MOA | 4600 | 50 | 40 | 24 | 1632 | 323 | 210 | 162 | 0.23 | 60 | 10 | 21 | 13 | 0 | 11 | 13 | 13 | 20 | 13 |
| NEA | 6100 | 60 | 86 | 55 | 1536 | 192 | 864 | 0 | 0.56 | 0 | 0 | 24 | 20 | 20 | 11 | 19 | 11 | 20 | 10 |
| NEB | 7200 | 55 | 70 | 43 | 1600 | 256 | 150 | 300 | 0.28 | 68 | 15 | 34 | 20 | 13 | 12 | 20 | 20 | 20 | 16 |

| Location | Tightness | Btu/DD/sf | Millions of Btu | Type of Heat | Comfort Level | Max. GH Temperature | Min. GH Temperature | |
|---|---|---|---|---|---|---|---|---|
| | | 0.96 | 12.00 | | | 90.0 | 40.0 | |
| | | 1.29 | 14.84 | | | 91.5 | 41.2 | |
| NYA | L | | | | | | | EARLY DEFICIENCIES CORRECTED |
| NYB | | 1.43 | 12.00 | W | N | 78.0 | 42.0 | NARROW SOUTH AIRFLOW SPACE |
| NYC | | 0.60 | 1.64 | E | | 90.0 | 38.0 | |
| PAA | | 0.83 | 13.00 | W | | 71.0 | 34.0 | FAN, ALL 4 SIDES BUFFERED ('HOLDOS') |
| PAB | | 5.90 | 92.00 | O | | 80.0 | 31.0 | UNFINISHED RETROFIT OF OLD FARMHOUSE |
| MDA | M | 1.92 | 24.00 | W | W | 96.0 | 38.0 | |
| MDB | M | 2.64 | 24.00 | W | N | 114.0 | 42.0 | |
| VAA | | 0.68 | 4.00 | W | | 130.0 | 41.0 | MONITORED BY UNIV. OF VA |
| VAB | | 0.66 | 6.00 | W | | 96.0 | 40.0 | TRIPLE GLAZED ON NORTH, EAST AND WEST |
| WVA | | | | E | | | | MODULAR ADAPTATION OF BEALE PLAN |
| WVB | | 1.15 | 18.00 | W | N | 90.0 | 44.0 | PRESSURE TREATED WOOD FOUNDATION |
| GAA | H | 1.13 | 6.00 | W | | 96.0 | 42.0 | MONITORED BY GEORGIA INST. TECHNOLOGY |
| TNA | | 3.97 | 12.00 | W | | | | LOG CABIN ('MARTIN') |
| TNB | M | 0.00 | 0.00 | O | | 110.0 | 30.0 | RETROFIT OF RANCH HOUSE |
| KYA | | 2.67 | 34.00 | W | | 75.0 | 55.0 | AIR CIRCULATES THROUGH LIVING SPACE |
| OHA | H | 0.78 | 8.00 | W | N | 90.0 | 45.0 | DAYTIME AIRFLOW THROUGH INTERIOR ('CLAYTON') |
| OHB | | 1.26 | 18.00 | W | | 90.0 | 55.0 | |
| OHC | | 0.92 | 18.00 | W | | 103.0 | 41.0 | |
| MIA | | 1.94 | 24.00 | W | | 100.0 | 35.0 | |
| MIB | | 2.45 | 50.00 | O | | 75.0 | 55.0 | FAN, PARTIAL ENVELOPE |
| MIC | | 3.54 | 50.00 | O | | 95.0 | 38.0 | LOOSE SOLARIUM SHADED BY PINES |
| IAA | | 0.54 | 6.50 | E | W | 95.0 | 40.0 | SIMILAR TO TOM SMITH HOUSE |
| IAB | | 0.25 | 5.12 | E | | 135.0 | 30.0 | FAN |
| IAC | | | | E | N | | | FAN, AIR FORCED THROUGH CONCRETE PLANK FLOOR |
| WIA | | 0.34 | 6.01 | E | W | 68.0 | 40.0 | UNOCCUPIED, FUEL USE PROJECTED FROM TEST PERIOD |
| MNA | | | | O | | 105.0 | 38.0 | ROOF INSULATION REMOVED PART OF WINTER |
| MNB | | 3.57 | 52.00 | W | | 77.0 | 38.0 | UNINSULATED INNER SHELL, MASS WALL, HUD MONITORED |
| MNC | | 1.58 | 27.30 | E | | 70.0 | 30.0 | 4 FT ROCKS OVER URETHANE INSULATION |
| SDA | | 0.24 | 4.30 | E | | 90.0 | 40.0 | FAN, AIRSHAFT, LOW GAIN |
| MTA | | | | E | C | 100.0 | 36.0 | 1250 GAL WATER STORAGE IN CRAWLSPACE |
| MTB | | 0.43 | 8.00 | W | | 100.0 | 33.0 | |
| ILA | | 0.11 | 0.57 | E | | 86.0 | 47.0 | |
| MOA | | 3.20 | 24.00 | W | | 75.0 | 40.0 | FAN, WOODSTOVE IN GREENHOUSE |
| NEA | | 0.75 | 7.00 | O | | 75.0 | 36.0 | 3-STORY VERTICAL GREENHOUSE GLASS ('NYHOLM') |
| NEB | | 0.15 | 1.71 | E | | 100.0 | 33.0 | INSULATED SHUTTERS ON SLOPING GLASS ('DEMMEL') |

| Location | Fahrenheit Degree-Days | Percent Sunshine | Owner Cost in Thousands of Dollars | Cost/Square Foot | Inner Floor Area | Greenhouse Area | Vertical South Glass | Sloped South Glass | Glass/Floor Ratio | Direct Glass | Other Buffered Glass | Buffer Height | R-Values of Insulations | | | | | | |
|---|---|---|---|---|---|---|---|---|---|---|---|---|---|---|---|---|---|---|---|
| | | | | | | | | | | | | | Ceiling | Floor | Inner Walls | Single Walls | Outer Walls | Roof | Foundation |
| MED.: | 6700 | 50 | 75 | 43 | 1800 | 300 | 228 | 170 | 0.23 | 48 | 24 | | 19 | 13 | 11 | 20 | 19 | 20 | 11 |
| MEAN: | 6400 | 49 | 80 | 44 | 1820 | 335 | 252 | 194 | 0.25 | 56 | 28 | | 17 | 13 | 12 | 22 | 20 | 24 | 12 |
| OKA | 3500 | 55 | 180 | 61 | 2950 | 1738 | 967 | 0 | 0.33 | 108 | 72 | 22 | 20 | 13 | 12 | 23 | 19 | 22 | 21 |
| COA | 5800 | 60 | 75 | 45 | 1655 | 320 | 262 | | 0.16 | 64 | 66 | 22 | 38 | | 15 | 20 | 19 | 38 | |
| COB | 6500 | 60 | 120 | 49 | 2420 | 520 | 366 | 108 | 0.20 | 99 | 62 | 20 | 13 | 11 | 13 | 22 | 22 | 19 | 8 |
| COC | 5800 | 60 | 45 | 25 | 1800 | 400 | 227 | 210 | 0.30 | 0 | 34 | 28 | 13 | 13 | 13 | 25 | 25 | 25 | 16 |
| COD | 6000 | 60 | 68 | 37 | 1800 | 360 | 257 | 193 | 0.25 | | | 24 | 12 | 18 | 12 | 25 | 25 | 18 | 16 |
| COE | 6000 | 60 | 62 | 35 | 1746 | 360 | 224 | 110 | 0.19 | 48 | 0 | 20 | 19 | 0 | 19 | 22 | 19 | 11 | 10 |
| COF | 6000 | 60 | 96 | 60 | 1600 | 392 | 320 | 216 | 0.34 | 64 | 0 | 20 | 11 | 11 | 11 | 33 | 27 | 19 | 15 |
| COG | 6000 | 60 | 105 | 33 | 3100 | 300 | 90 | 210 | 0.10 | 15 | 36 | 22 | 2 | 2 | 2 | 28 | 28 | 45 | 10 |
| COH | 6000 | 60 | 55 | 32 | 1708 | 400 | 242 | 222 | 0.27 | 88 | 0 | 21 | 11 | 11 | 11 | 19 | 19 | 30 | 10 |
| COI | 6000 | 60 | 51 | 39 | 1296 | 240 | 141 | 211 | 0.27 | 77 | 0 | 22 | 11 | 0 | 11 | 19 | 19 | 30 | 8 |
| COJ | 10000 | 50 | 100 | 53 | 1870 | 400 | 194 | 230 | 0.23 | 18 | 0 | 24 | 11 | 11 | 11 | 19 | 19 | 22 | 19 |
| COK | 6000 | 50 | 59 | 38 | 1536 | 256 | 450 | 0 | 0.29 | 72 | 0 | 34 | | | | | | | |
| WYA | 8260 | 75 | 140 | 64 | 2175 | 227 | 312 | 170 | 0.22 | 34 | 9 | 25 | 19 | 2 | 11 | 25 | 25 | 19 | 11 |
| WYB | 8260 | 75 | 58 | 36 | 1590 | 266 | 288 | 140 | 0.27 | 23 | 6 | 25 | 16 | 2 | 11 | 25 | 25 | 16 | 19 |
| WYC | 9300 | 75 | 30 | 25 | 1173 | 208 | 175 | 130 | 0.26 | 24 | 21 | 22 | 16 | 2 | 11 | 25 | 25 | 16 | 11 |
| WYD | 8500 | 75 | 55 | 30 | 1794 | 307 | 291 | 218 | 0.28 | 26 | 4 | 23 | 19 | 2 | 11 | 25 | 25 | 19 | 11 |
| WYE | 7700 | 75 | 50 | 30 | 1643 | 288 | 315 | 48 | 0.22 | 15 | 4 | 25 | 16 | 2 | 11 | 25 | 25 | 16 | 11 |
| IDA | 7000 | 40 | 75 | 41 | 1792 | 320 | 210 | 125 | 0.19 | 30 | 15 | 23 | 7 | 7 | 12 | 19 | 19 | 7 | 5 |
| CAA | 3000 | 60 | 130 | 68 | 1900 | 484 | 550 | 0 | 0.29 | | 0 | 22 | 22 | 1 | 13 | 22 | 13 | 19 | 15 |
| CAB | 8200 | 55 | 75 | 68 | 1092 | 356 | 158 | 223 | 0.35 | 145 | 94 | 31 | 11 | 11 | 11 | 19 | 19 | 30 | 10 |
| CAC | 3000 | 50 | 30 | 18 | 1650 | 240 | 180 | 144 | 0.20 | 110 | 0 | 22 | 19 | | 11 | 19 | 19 | 30 | 11 |
| CAD | 3000 | 50 | 30 | 31 | 960 | 448 | 324 | | 0.34 | 174 | 0 | 18 | 19 | 19 | 11 | 19 | 19 | 30 | 11 |
| CAE | 4600 | 45 | | | | 320 | 255 | 114 | | | | 23 | 19 | 0 | 11 | 19 | 13 | 19 | 18 |
| CAF | 4600 | 45 | 59 | 43 | 1352 | 260 | 150 | 100 | 0.18 | 100 | 0 | 26 | 19 | | 11 | 19 | 19 | 50 | 15 |
| CAG | 5000 | 75 | | | 1620 | 300 | 163 | 113 | 0.17 | 110 | 0 | | 11 | 0 | 11 | 19 | 19 | 19 | 8 |
| CAH | 8100 | 60 | | | 1700 | 360 | 260 | 200 | 0.27 | 90 | 0 | 24 | 13 | 13 | 13 | 19 | 19 | 19 | 13 |
| CAI | 8150 | 60 | 54 | 30 | 1800 | 250 | 235 | 155 | 0.22 | 100 | 0 | 24 | 13 | 13 | 13 | 19 | 19 | 19 | 13 |
| CAJ | 3000 | 50 | 98 | 40 | 2400 | 418 | 0 | 440 | 0.18 | 30 | 38 | 21 | 13 | 0 | 0 | 19 | 19 | 19 | 0 |
| ORA | 5000 | 40 | | | 1920 | 480 | 195 | 180 | 0.20 | 70 | 0 | | 11 | 11 | 11 | 23 | 19 | 20 | 10 |
| CDA | 7200 | 35 | 65 | 30 | 2110 | 432 | 200 | 200 | 0.19 | | | 26 | 28 | 28 | 11 | 28 | 33 | 28 | 15 |
| CDB | 8000 | 30 | 125 | 64 | 1950 | 400 | 0 | 777 | 0.40 | 90 | 40 | 24 | 0 | 7 | 7 | 20 | 20 | 10 | 10 |
| CDC | 7200 | 30 | | | 1000 | 200 | | | 0.25 | | | 30 | 12 | 12 | 12 | 27 | 27 | 40 | 20 |
| CDD | 8800 | 30 | | | 2048 | 287 | 240 | 131 | 0.18 | 96 | 18 | 30 | 12 | 12 | 12 | 27 | 27 | 27 | 20 |

| Location | Tightness | Btu/DD/sf | Millions of Btu | Type of Heat | Comfort Level | Max. GH Temperature | Min. GH Temperature | |
|----------|-----------|-----------|-----------------|--------------|---------------|---------------------|---------------------|---|
| | | 0.96 | 12.00 | | | 90.0 | 40.0 | |
| | | 1.29 | 14.84 | | | 91.5 | 41.2 | |
| OKA | | 0.53 | 5.44 | E | W | 88.0 | 52.0 | SWIMMING POOL IN SUNSPACE |
| COA | M | 0.42 | 4.00 | W | | 86.0 | 48.0 | AIRFLOW THROUGH 2ND STORY ROOMS ('WILLIAMS') |
| COB | | 0.38 | 6.00 | E | W | | | EXPOSED WINDY SITE |
| COC | | 0.57 | 6.00 | W | | 105.0 | 45.0 | FAN, 54 WATER DRUMS IN BASEMENT, INS. UNDER SLAB |
| COD | | 0.96 | 10.40 | W | | 95.0 | 40.0 | |
| COE | H | 0.57 | 6.00 | W | | 90.0 | 45.0 | "INNER HOUSE NEVER ABOVE 75 OR BELOW 60" |
| COF | | 0.17 | 1.68 | E | N | 95.0 | 50.0 | |
| COG | | 0.32 | 6.00 | W | | 80.0 | 50.0 | FAN-FORCED REVERSE LOOP, R-14 SHUTTERS ON SOUTH |
| COH | M | 0.05 | 0.50 | W | | 102.0 | 51.0 | |
| COI | H | 0.26 | 2.00 | W | N | 104.0 | 54.0 | |
| COJ | H | | | E | | | | FAN |
| COK | L | 0.00 | 0.00 | O | C | | | NO AUXILIARY HEAT USED |
| WYA | | | | W | N | | | |
| WYB | | 0.24 | 3.20 | W | N | 146.0 | 46.0 | |
| WYC | | | | W | N | | | |
| WYD | | | | W | N | | | |
| WYE | | | | W | N | | | |
| IDA | | 0.96 | 12.00 | W | | 85.0 | 50.0 | |
| CAA | H | 0.19 | 1.10 | W | N | 120.0 | | |
| CAB | | 1.44 | 12.90 | W | N | 115.0 | 39.0 | FAN |
| CAC | | 0.00 | 0.00 | | | 125.0 | 40.0 | FAN, 15" PIPES FOR AIRFLOWS UNDER SLAB, NO AUX. |
| CAD | | 0.42 | 1.20 | E | | 82.0 | 52.0 | AIRFLOW THROUGH CLOSETS, STAIRS ('STEWART') |
| CAE | | | | W | | | | FAN |
| CAF | | 0.96 | 6.00 | W | | 80.0 | 55.0 | |
| CAG | L | | | W | | 80.0 | 44.0 | |
| CAH | | | | | | | | FAN |
| CAI | M | 0.55 | 8.00 | W | N | 80.0 | 45.0 | TOM SMITH HOUSE |
| CAJ | | 2.50 | 18.00 | W | | 120.0 | 50.0 | FAN, SINGLE-GLAZED, DAMPERS IN LOOP ('ANAWALT') |
| ORA | | 0.42 | 4.00 | W | | 84.0 | 50.0 | |
| CDA | | 0.75 | 10.30 | O | | | 28.0 | FAN, SHUTTERED EAST & WEST GLASS ('ALPHA') |
| CDB | L | 2.56 | 40.00 | W | | 95.0 | 37.0 | FAN, ISOLATED ROCK STORAGE, INTERNAL MASS, FPLC. |
| CDC | | 2.78 | 20.00 | W | W | 90.0 | 53.0 | PRESSURE-TREATED WOOD FOUNDATION, AIRTIGHT |
| CDD | | 2.41 | 43.50 | W | | 80.0 | 42.0 | ANALYZED FOR NAT'L RESEARCH COUNCIL CANADA |

### Climate

The average number of degree-days for the houses in the survey was about 6,500. There were two houses in 10,000 degree-day climates and several in the 3,000-4,000 range. The greatest number of houses in the survey are located in the Northeast, about one-third in New England.

The average percent of possible sunshine for the houses was 50 percent, with several in the 30-40 percent range and several in Colorado and Wyoming at 60 percent or more.

### Auxiliary Heating

We were able to get information about yearly auxiliary heat requirements from 90 of the respondents. The figure for auxiliary heat represents the total amount of energy used in one year specifically for keeping the house warm, and does not include such intrinsic energy as heat from occupants or the electricity used to operate lights, cooking stoves, or appliances.

Three of the respondents reported needing no auxiliary heat at all, more than half used the equivalent of one cord of wood or less, and 85 percent used the equivalent of less than two cords. The mean amount of auxiliary heat used was 14.8 million Btus, which represents just over one cord of hardwood, or $304 for electric baseboard heat at $.07 per kWhr.

Most of the houses used woodstoves for heat, and the next largest group used electric heat.

The following chart shows the distribution of the houses by yearly auxiliary heat requirements. In this chart, as in later charts in this chapter, each "x" on the right represents one house that had a value in the range indicated on the left.

### Million Btus Yearly Auxiliary Heat

| | | | |
|---|---|---|---|
| 0.0 | — | 4.9 | xxxxxxxxxxxxxxxxxx |
| 5.0 | — | 9.9 | xxxxxxxxxxxxxxxxxxxx |
| 10.0 | — | 14.9 | xxxxxxxxxxxx |
| 15.0 | — | 19.9 | xxxxxxxxxxxx |
| 20.0 | — | 24.9 | xxxxxxxxxxxxx |
| 25.0 | — | 29.9 | xx |
| 30.0 | — | 34.9 | xx |
| 35.0 | — | 39.9 | xx |
| 40.0 | — | 44.9 | xx |
| 45.0 | — | over | xxx |

At $.07 per kilowatt-hour for electric resistance
heat, each 5 million Btus in the chart above
represents an additional yearly heating cost of
about $100.

## Btu/DD/sf

Some of the most impressive results from this
survey are the figures obtained for auxiliary
heating, converted to Btu per degree-day per square
foot. Btu/DD/sf is a unit of measurement which is
frequently used as an overall indicator of a
building's energy efficiency, for comparing the
performance of houses of different sizes and in
different locations. It is obtained by dividing the
net yearly heat usage (in Btus) by the number of
heating degree-days for the location, and then
dividing by the square footage of floor area in the
house.

As a unit of measurement, Btu/DD/sf can be
misleading, since it does not take into account
some important factors. For example:

— A house in a cold climate will have a higher
Btu/DD/sf than an identical house in a warmer
climate. This is probably the biggest flaw in the use
of Btu/DD/sf as a unit of comparison. The total
heat loss from a house is directly proportional to
the temperature difference between the inside and
outdoors and is therefore proportional to the
number of degree-days of the location. However,
both internal gains and solar gains are unrelated to
degree-days, so in a warmer climate those gains
would provide a relatively greater proportion of the
total heat loss than identical solar and internal
gains would provide in a colder climate. In a
superinsulated house, if internal gains provided 1/4
of the total heat needs in a cold climate, the same
amount of internal gains might provide 3/4 or
perhaps all of the heat needs of the same house in
a moderate climate.

— The percent of sunshine available in a location
has a large effect on the performance of any house
depending much on solar energy.

— Larger buildings tend to require less heating
energy per square foot of floor space than smaller
buildings.

Of the 90 houses for which we had information on
auxiliary heat use, 48 houses, or 53 percent, used

less than 1 Btu/DD/sf. This is an extremely low heating requirement, equivalent to less than one cord of wood a year for a 1500sf house in a 6500 degree-day climate.

For comparison with other kinds of houses, their normally expectable heating requirements, in Btu/DD/sf, are:

| | |
|---|---|
| 1976 inventory of US housing | c. 15 Btu/DD/sf |
| Northeast Solar Energy Center 1980-1981 study of solar houses | c. 8 Btu/DD/sf |
| Houses described in the HUD/DOE "First Passive Solar Home Awards" | c. 4 Btu/DD/sf |

The floor area used here in computing Btu/DD/sf specifically did not include the area of the sunspace or any other unheated locations (although in a few of the houses some heat was supplied to the sunspace or other buffering spaces). If the sunspace had been included in the floor area as it often is, the resultant figures would have been even lower and more impressive than they already are.

The chart below shows the distribution of the 90 houses by Btu/DD/sf for auxiliary heating.

**Btu/DD/sf**

| | | | |
|---|---|---|---|
| 0.00 | — | 0.24 | xxxxxxxxxxx |
| 0.25 | — | 0.49 | xxxxxxxxxxx |
| 0.50 | — | 0.74 | xxxxxxxxxxxxxx |
| 0.75 | — | 0.99 | xxxxxxxxxxxx |
| 1.00 | — | 1.24 | xxxxxxxx |
| 1.25 | — | 1.49 | xxxxxx |
| 1.50 | — | 1.74 | xxxxx |
| 1.75 | — | 1.99 | xx |
| 2.00 | — | 2.24 | xxxx |
| 2.25 | — | 2.49 | xxx |
| 2.50 | — | 2.74 | xxxx |
| 2.75 | — | 2.99 | xxx |
| 3.00 | — | 3.24 | xxx |
| 3.25 | — | 3.49 | |
| 3.50 | — | 3.74 | x |
| 3.75 | — | 3.99 | x |
| 4.00 | — | 4.24 | |
| 4.25 | — | 4.49 | x |
| 4.50 | — | 4.74 | |
| 4.75 | — | 4.99 | |
| 5.00 | — | 5.24 | |
| 5.25 | — | 5.49 | |
| 5.50 | — | 5.74 | |
| 5.75 | — | 6.00 | x |

In many of the houses it is possible to see reasons for unusually high or low fuel consumption by examining the data on the house; for example, poor or incomplete insulation is clearly a factor in some of the houses with higher fuel use (the house with the highest fuel consumption on the graph was an unfinished retrofit of an old leaky farmhouse). In some cases it is necessary to look at several different elements to see why a particular house performed as it did, and in a few cases the key factor seems to be something not identifiable in the numbers; there are houses near both the high and low ends of the Btu/DD/sf chart which just do not seem to belong where they are.

There was some correlation between the percent of sun and the Btu/DD/sf figure. Many of the houses with the lowest auxiliary heat use are in areas that have a lot of winter sunshine (around 60 percent of possible sun), and many of the houses that used a comparatively large amount of purchased heat were in areas with only 30-40 percent of possible winter sunshine. This simply confirms the fact that these houses are, after all, solar houses.

Unexplainably high or low fuel consumption is probably due partly to the lack of certain data about the tightness of construction and the comfort levels preferred by occupants, although we tried to determine these factors.

We do not know much about the occupants' lifestyles. The use of heat-generating appliances varies with the size of the family, the number of appliances, the frequency of use, and their efficiency. If internal heat sources typically produce an average of about 50,000 Btu per day, doubling that amount over a four month heating season would make a difference equivalent to about 1/2 cord of firewood, or about .5 Btu/DD/sf for the average of this survey.

## Cost

The costs of the houses in the survey varied widely. About 10 were in the $30,000 range, and there were two houses over $180,000. In many cases the owners supplied much of the labor, the value of which was not always included in the reported cost. The average cost was around $80,000.

### Owner's cost, excluding site, design, appliances

| | | |
|---|---|---|
| Under | $25,000 | xx |
| 25,000 — | 49,999 | xxxxxxxxxxxxx |
| 50,000 — | 74,999 | xxxxxxxxxxxxxxxxxxxxxxxxxxxxxxxxxx |
| 75,000 — | 99,999 | xxxxxxxxxxxxxxxxxxxx |
| 100,000 — | 124,999 | xxxxxxxxxxxxxxx |
| 125,000 — | 149,999 | xxxxxxxx |
| 150,000 — | over | xxxx |

On a square foot basis, the average cost was $44/sf. Again the range is large, from under $20/sf to over $70/sf. In computing the cost per square foot, the floor area was calculated as noted above: the greenhouse/sunspace area was not included. This makes the cost appear higher than it normally would, since the greenhouse in almost all the houses does provide useful space. So the figures for cost per square foot would seem to refute the argument that an envelope house necessarily costs more per square foot of inner house than other types of designs. However, as noted above, a fairly large proportion of these houses were built with at least some of the owners' labor, which affects the cost figures to an unknown degree.

### Cost per square foot of inner house

| | | |
|---|---|---|
| Under | $20 | xx |
| 20 — | 29 | xxxxxxxxxxxx |
| 30 — | 39 | xxxxxxxxxxxxxxxxxxxxxxxxxxx |
| 40 — | 49 | xxxxxxxxxxxxxxxxxxxxxxx |
| 50 — | 59 | xxxxxxxxxxxxxxxxxx |
| 60 — | 69 | xxxxxxxxxx |
| 70 — | 80 | xxx |

## Glass

Most of the houses had a fairly generous south glazing area, with an average of over 400sf and an average ratio of south glazing to interior floor area of .25.

### South glass total area, square feet

```
Under       100
100  —  199   xxxxxxx
200  —  299   xxxxxxxxxxxx
300  —  399   xxxxxxxxxxxxxxxxxxxxxxxxxxxxxx
400  —  499   xxxxxxxxxxxxxxxxxxxxxxxxxxxx
500  —  599   xxxxxxxxxxxxxxxx
600  —  699   xxxxxx
700  —  799   x
800  —  899   xxxx
900  — 1000   xxx
```

### Ratio of south glass to total inner house floor area

```
Under      .10   xxx
.10  —  .19   xxxxxxxxxxxxxxxxxxxxxxxxxxxxxxxxxxxxx
.20  —  .29   xxxxxxxxxxxxxxxxxxxxxxxxxxxxxxxxxxxxxxxxxxxxxxx
.30  —  .39   xxxxxxxxxxxxxx
.40  —  .49   xxx
.50  —  .59   xxxx
.60  —  .70   x
```

Most of the houses had fairly conservative amounts of glass on other sides than south. The average total window area on the east, north and west was 84sf, or 4.6 percent of the floor area. Most windows were double-glazed, with 23 percent of the east and west windows triple-glazed, and 15 percent of the north windows triple-glazed.

## Greenhouse Temperatures

Along with the net auxiliary heating load, temperature fluctuations in the greenhouse or sunspace provide one of the most important indications of envelope house performance. For many people, the plant-growing capacity of the greenhouse is a large part of its value, and the more reliably it maintains good growing temperatures the more valuable it is. In addition, severe overheating in a sunspace on sunny days is an indication of inefficient use of collected solar energy, because whenever the sunspace is overheated its losses through large windows are increased (because of the increase in temperature difference to outdoors). An overheated sunspace means that a lot of the collected heat is not getting into storage. It also means that the space is not a comfortable place for people.

Of the 92 houses for which we had information on minimum greenhouse temperature, 12 got as cold as 32F or lower; at least three of those were unfinished and not tightly closed in at the time. The average minimum temperature in the greenhouse was about 40F.

### Minimum Greenhouse Temperatures

```
24  —  27  xx
28  —  31  xxxxxxxxx
32  —  35  xxxxxxxxxx
36  —  39  xxxxxxxxxxxxxxxxxxx
40  —  43  xxxxxxxxxxxxxxxxxxxxxx
44  —  47  xxxxxxxxxxxxx
48  —  up  xxxxxxxxxxxxxxxxx
```

Most of the houses had a maximum greenhouse temperature between 80 and 100F. Two of the houses stayed below 70F in the winter and two got above 140F.

### Maximum Winter Greenhouse Temperature

```
 60  —   69  xx
 70  —   79  xxxxxxxx
 80  —   89  xxxxxxxxxxxxxxxxxxxxxxxxxxxxxx
 90  —   99  xxxxxxxxxxxxxxxxxxxxxxxxxxx
100  —  109  xxxxxxxxxxxx
110  —  119  xxxxxxx
120  —  129  xxxx
130  —  139  xx
140  —  150  xx
```

## Insulation

If the insulation of the outer and inner shells of the survey houses were combined in a single shell (as in a conventional structure) the average total R-values would be:

walls:          R-26
roof/ceiling:   R-39
floor:          R-13

Since by far the largest window area in these houses is on the south and most windows were double-glazed, the average combined glazing thickness is very nearly four layers.

A surprising 23 houses (or 23 percent of the sample) had no insulation in the floor (between the inner house and the crawlspace or basement). Almost all the remaining houses had 4 to 6 inches of floor insulation.

In general, the outer shells were better insulated than the inner shells. The average outer wall R-value was 20, whereas the average inner wall R-value was only 12.

85 percent of the houses had R-10 to R-20 ceiling insulation (between the inner house and the attic airspace). The roofs (between the attic airspace and outdoors) were better insulated, with more than half of the respondents using around R-20 and 34 percent using R-30 or more.

12 of the houses had some insulation between the crawlspace or basement and the deep earth, either at or near the surface. This was usually around R-10. In those houses, the deep earth mass would presumably be less effective in preventing the greenhouse from freezing, but out of the 10 for which we had temperature data only 2 reached freezing or below, which is 20 percent of this sample as compared with 13 percent for all the houses.

## Fans

Quite a few houses (23) used some sort of fan system for moving greenhouse air. Their typical performance is not noticeably different from houses that relied solely on natural convection.

# Reference Section
## A
## ABBREVIATIONS

| | |
|---|---|
| A | area |
| AC/hr | air changes per hour |
| BLC | building load coefficient |
| Btu | British thermal unit |
| C | conductance |
| cf | cubic feet |
| cfm | cubic feet per minute |
| DD | degree-day |
| deg | degree |
| DHW | domestic hot water |
| EPS | expanded polystyrene "beadboard" |
| F | Fahrenheit |
| fpm | (linear) feet per minute |
| fps | feet per second |
| ft | feet |
| kWhr | kilowatt-hour |
| LCR | Load Collector Ratio |
| MMBtu | million Btu |
| PCM | phase change material |
| R | resistance |
| RH | relative humidity |
| sf | square feet |
| SLR | Solar Load Ratio |
| T | temperature |
| U | conductance |
| UA | conductance (U) times area (A); or area divided by R-value |
| W | watt |

**Reference Section**
**B**
**GLOSSARY**

**absorptance:** The fraction of radiant energy which is absorbed by a material, rather than being transmitted through it or reflected by it. The absorptance of a material corresponds roughly to the darkness of the surface. Examples of approximate values for solar absorptance:

| | |
|---|---|
| polished aluminum | .1 |
| white surface | .2 to .4 |
| green, red, or brown | .5 to .7 |
| blue surface | .8 |
| black surface | .9 |
| snow | .3 |
| pine | .6 |
| concrete | .6 |
| asphalt shingles | .8 |
| sand | .8 |

**active system:** An indirect gain solar heating system in which fans or pumps are used to transfer heat, in fluids such as air or water, into and out of storage.

**air change rate:** Air changes per hour (AC/hr) is used as a measure of the infiltration and fresh air intake for a house. It is the number of times in each hour that a volume of air equivalent to the volume of the house enters from outdoors (and the same volume of air leaves the house). An old leaky house could have an air change rate of more than 6 AC/hr, whereas a very tight new superinsulated house might have a natural air change rate of 0.1 AC/hr or even less.

**airlock:** A cold-climate entrance space with two tight doors, one to the house and one to the outside, so one door can be closed when the other is open to minimize the amount of cold exterior air that will infiltrate the house when people enter or leave. The airlock is less necessary in tight houses that have little infiltration, because for much cold air to come in through an open doorway, warm air must have places where it can go out.

**altitude:** The angle of the sun above the theoretical horizon.

**ambient temperature:** The surrounding outdoor temperature.

**aperture:** An opening. Usually refers to an area of south glazing: the area that the sun can shine through.

**ASHRAE:** American Society of Heating, Refrigerating, and Air-Conditioning Engineers, or the Society's Handbooks of technical information — particularly "1981 Fundamentals."

**auxiliary heat:** Heat which is supplied specifically for keeping a house warm, from a furnace, woodstove, etc. Does not include heat from appliances, etc. (internal gains). Yearly auxiliary heat use is frequently measured in units of millions of Btu (MMBtu). The auxiliary heat required by a house in a cold climate could range from well over 100 MMBtu/yr for an older, poorly insulated house to less than 5 MMBtu/yr for a new superinsulated or buffered house.

**azimuth:** The angle of the sun to the east or west of due south. The sun's azimuth is zero at solar noon.

**berming:** The placement of earth against one or more of the walls of a house (especially the north wall). The earth provides protection against hot and cold weather extremes and wind. Berming can be achieved either by setting the house into the earth (ideally, into a south-facing slope) or by piling up earth against one or more sides.

**Btu:** British thermal unit. This is the basic conventional unit of heat energy. It is the amount of energy required to heat one pound of water one degree Fahrenheit. Examples of typical heat output in Btu's:

| | |
|---|---|
| one person's body heat: | 300 Btu/hr |
| 100-watt light bulb: | 341 Btu/hr |
| noon winter sun, 1sf S window: | 200 Btu/hr |
| conventional home furnace: | 120,000 Btu/hr |
| medium-sized woodstove: | 40,000 Btu/hr |

**building load coefficient (BLC):** The overall heat loss rate of a building, in Btu/hr/deg F or Btu/DD.

**capacitance, capacity:** The measure of an object's ability to store heat. Conventionally in units of Btu/deg F, or (per cubic foot) Btu/cf/deg F. Capacitance determines how slowly an object changes temperature when it gains or loses an amount of heat.

| material | capacitance (Btu/cf/deg F) |
| --- | --- |
| water | 62 |
| concrete | 25 |
| brick | 24 |
| sand | 18 |
| gypsum | 15 |
| hardwood, dried | 13 |
| softwood, dried | 11 |
| fiberglass insulation | 0.2 |
| air | 0.018 |

**clerestory:** A vertical window in a roof.

**conductance:** The ability of a material to conduct heat. The opposite of resistance (R-value), which is the ability of a material to impede the flow of heat. Conventionally in units of Btu/hr/deg F, or (as a per square foot measure) Btu/hr/sf/deg F.

**conduction:** The flow of heat directly through a material.

**conservation:** Refers to measures which slow the flow of heat from a house, usually by adding insulation or reducing air infiltration.

**convection:** The transfer of heat by moving air, water, or other fluid.

**degree-day (DD):** Heating degree-days are used as a measure of the coldness of a climate, for estimating auxiliary heat use. For each day, the number of degree-days is equal to the difference between the average temperature and 65F, if the average is below 65F. (Heating degree-days are normally computed with a base of 65F.) For example, one day's average temperature of 30F is equivalent to 35 degree-days. Monthly and yearly degree-day totals and averages are compiled by many weather stations, and are available in many books or from local weather stations. The heating needs of modern efficient houses are frequently not directly proportional to the number of degree-days of the location.

**Delta T:** Difference in temperature. The amount of heat conducted through a material is proportional to the temperature difference, or Delta T, between the two surfaces of the material.

**desiccant:** A moisture absorber.

**dew point:** As a volume of air is cooled, it can hold less and less moisture. The dew point is the temperature at which air containing a particular amount of moisture is cold enough that it can hold no more water — the air is saturated. When further cooled, some of the water vapor condenses to form liquid water (or, if below 32F, frost).

**direct gain:** A solar heating strategy in which sunshine is received directly into the living space, and some of the sun's heat is absorbed there by the house and its contents and by added heat storage materials such as a concrete floor.

**diurnal storage:** Daily heat storage, as when heat is absorbed in a thermal storage mass during the day and released at night.

**emittance:** The fraction of radiant energy given off (emitted) by a material, as compared with the maximum amount that could be emitted by an ideal "blackbody" at the same temperature.

**envelope:** In architecture, the building shell. The word has recently referred also to the "envelope" of air surrounding much of the occupied space of the double-shell "envelope" house.

**forced convection:** The transfer of heat using fans or pumps to move a fluid such as air or water.

**glazing:** Glass or other transparent or translucent materials used to enclose a space while allowing light through. Single, double, triple, and quadruple are used to refer to the number of glazing layers; it is the number of layers that principally determines the thermal resistance (R-value) of glazing.

**gravity airflow:** Same as natural convective airflow.

**heat exchanger:** A device which transfers heat from one fluid to another. For example, a coil of pipe immersed in a tank of water can transfer heat from water in the pipe to the water in the tank without the two bodies of water mixing. An air-to-air heat exchanger is a device which uses the heat from exhaust air from a house to preheat fresh incoming air from outdoors. It provides a means of obtaining fresh air in a tight house with a minimal heat loss.

**heat, low grade:** Refers to heat that is lower in temperature than that required for human comfort but warm enough to keep a greenhouse above freezing.

**indirect gain:** A solar heating system in which the solar collector is separated from the living space of the house. Trombe walls and sunspaces are examples of indirect gain devices.

**infiltration:** Outside air entering a house, through cracks, etc. Usually described in terms of air changes per hour (AC/hr). Infiltration heat losses typically account for 1/4 to 1/2 of the total losses from a house.

**insolation:** Stands for INcident SOLar radiATION. The amount of solar gain striking a surface during a specified period.

**internal gains, intrinsic gains:** Heat generated in the normal occupancy of a house, including heat from appliances, bodies, cooking stoves, hot water, etc. It does not include heat which is supplied specifically for heating a house, such as from a furnace or woodstove, and does not include solar heat gain. The amount of internal gains generated depends on many factors such as the number of occupants, the amount of time when occupants are at home, and the number and type of appliances and how much they are used. Estimates of typical amounts of internal gains range from below 10,000 to above 30,000 Btu per occupant per day. The heat produced per hour by most electrical appliances (when being used) can be calculated as the operating wattage times 3.413, assuming that the energy produced is released in the house. The major common source of heat which often does not stay inside long enough to release its heat is the hot water which goes down the drain and out of the house before it cools to room temperature.

**latent heat:** Heat stored when a material changes state, from a solid to a liquid or from a liquid to a gas. For example, the evaporation of liquid water into water vapor requires a certain amount of heat energy; one form of latent heat is the heat which is released when that water vapor condenses back into liquid. Humid air which contains a lot of water vapor contains much more latent heat than dry air. Latent heat storage, unlike sensible heat storage, does not involve changing the temperature of the storing material. See also phase change material.

**movable insulation:** Insulation that can be applied at a window, to reduce the heat flow through the window at chosen times (usually at night). Some forms of movable insulation are: insulating shades or shutters, roll-down window quilts, rigid foam pieces that are placed manually, and panels that slide into window openings from the side. Since the R-value of double-glazing is only about R-2, the heat loss from large window areas (which may collect much solar heat during the day) can be markedly reduced at night by the use of movable insulation.

**natural convection:** In a fluid, for example air, if one part is warmer than another, the warmer, lighter portion will tend to rise while the cooler, heavier portion sinks. Natural convection is the transfer of heat by the resultant movement of fluid. Some products of natural convection are stratification, the convective loop in an envelope house, and thermosiphoning.

**night insulation:** Same as movable insulation.

**passive system:** A solar heating system which relies solely on natural processes to move heat, as compared with active systems which rely on fans or pumps.

**permeance:** The measurement of how easily water vapor can pass through a material, conventionally in units of perms. For example:

| | |
|---|---|
| 6-mil polyethylene | .06 perms |
| 1/4" exterior plywood | .7 perms |
| 4" concrete | .8 perms |
| 1" extruded polystyrene | 1 perm |
| 1" expanded polystyrene | 2-5 perms |
| 3/8" gypsum | 50 perms |
| Tyvek infilt. barrier | 94 perms |

**phase change material (PCM):** A material which stores much heat by changing state, from a solid to a liquid. For example, some eutectic salts can store about 100 Btu/lb when heated above their melting point, which is usually near 90F. Those 100 Btu's of latent heat are released when it solidifies ("freezes") again. The fact that the melting point of these salts is above the human comfort range (unlike ice/water), but not too much above (unlike most solids), makes them useful for storing solar heat for houses.

**photovoltaics:** Cells which convert solar radiation into electricity. With improvements in technology and larger-scale production, the cells are becoming less expensive and more cost-effective.

**plenum:** An air chamber or passageway, for example the airspace above the ceiling or between two north walls in an envelope house.

**radiation:** The transfer of energy through space, directly from one object to another. Radiation is transferred as electromagnetic waves, including: infrared radiation, visible light, ultraviolet radiation, and radio waves. Solar radiation contains radiation in all those different wavelengths, although by the time it reaches the earth's surface about half of it is in the range of visible light.

**reflectance:** The fraction of radiant energy reflected from a surface, rather than being absorbed by or transmitted through the material. See also absorptance.

**relative humidity (RH):** The amount of water vapor present in air, expressed as a percentage of the maximum amount that air can hold at the given temperature. As a body of air is cooled, the maximum amount of vapor that it can hold is reduced, so the relative humidity increases. If cooled to the point where it cannot hold any more water (the dew point, 100% RH), some of the water vapor will condense to form liquid water.

**resistance:** The ability of a material to resist the flow of heat through it; the opposite of conductance. See R-value.

**R-value:** The measure of a material's ability to resist the flow of heat through it; the measure of its insulating effectiveness. R-value is the inverse of U-value. Some typical R-values are:

| | |
|---|---|
| 3-1/2″ fiberglass batts | 11 to 13 |
| 5-1/2″ fiberglass batts | 19 |
| 1″ cellulose | 3.5 |
| 1″ expanded polystyrene | 4 |
| 1″ extruded polystyrene | 5 |
| 1″ urethane | 6 |
| 1″ polyisocyanurate | 7 |
| 1″ airspace, one reflective surface | 2 to 3 |
| 1″ airspace, both dull surfaces | 1 |
| 1″ pine board | 1.25 |
| 1/2″ gypsum | .45 |
| 8″ concrete blocks | 1.1 |
| 1/2″ plywood | .6 |
| single-layer glass | .9 |
| double-glazing | 1.8 |
| triple-glazing | 2.7 |
| quadruple-glazing | 3.6 |
| inside air film | .68 |
| outside air film | .17 |

**sensible heat:** Heat which is stored in a substance by raising its temperature, as opposed to latent heat, which can be stored or released without changing the storage material's temperature.

**Solar Load Ratio:** Refers to a method of calculating the performance of a solar house using correlations based on many computer simulations. Described in "Passive Solar Design Handbook," Volumes 2 and 3. "Solar Heating Fraction," "Solar Savings Fraction," and "Load Collector Ratio" are other terms from those books.

**specific heat:** The heat storage ability of a material, per pound of mass. The storage ability per cubic foot (the volumetric capacity) is determined by multiplying the material's specific heat times its density (pounds/cubic foot). Specific heat is conventionally measured in units of Btu/lb/deg F.

**stratification:** The process of air (or other fluid) at different temperatures forming layers, hotter at the top and cooler below.

**superinsulation:** The system of reducing heat losses from a house to such an extent that very little added heat is required to maintain the occupants' comfort.

**thermal mass:** Materials and combinations of materials which are effective in storing heat, for example the enormous thermal mass of the earth underneath a house, or the thermal mass of a masonry wall.

**thermosiphoning:** The movement of a fluid such as air or water in a convective loop. Heat applied to one side of the loop makes the fluid lighter on that side, causing it to rise while the cooler and heavier fluid on the other side of the loop falls and moves across to replace it. Thermosiphoning is used in many passive solar heating systems, such as domestic hot water collectors, air heaters for space heating, envelope houses using natural convection, and vented Trombe walls.

**transmittance:** The extent to which a material allows radiation to pass through it. A transparent or translucent material is one which transmits much visible light.

**U-value:** The measure of conductance; the inverse of R-value.

**vapor barrier:** A material with very low permeance used to slow the movement of water vapor into or through a structure. Good super-insulated construction requires the installation of a continous vapor barrier near the warm side (interior) of the building shell.

## Reference Section
## C
## RELATED READING

### General

Alberta Agriculture, Home and Community Design Branch. *Low Energy Home Designs.* Alberta Agriculture, 9718 — 107 Street, Edmonton, Alberta T5K 2C8. Clear presentation of some design and construction details, with insightful comments about house plan specifics.

Anderson, Bruce with Riordan, Michael. *The Solar Home Book.* Brick House. Introduced a lot of people to the theory and practice of solar heating.

Anderson, Bruce and Wells, Malcolm. *Passive Solar Energy.* Brick House. Principles of the major approaches to passive solar design, with a section on passive cooling.

*ASHRAE Handbook: 1981 Fundamentals.* American Society of Heating, Refrigerating, and Air-Conditioning Engineers, 1791 Tullie Circle NE, Atlanta, GA 30329. Very technical reference for almost all aspects of heating and cooling.

Cole, John N and Wing, Charles. *From The Ground Up.* Atlantic; Little, Brown. Useful perspectives and information, especially for first-time home builders.

Kern, Ken. *The Owner-Built Home.* Owner-Builder Publications, Box 817, North Fork, CA 93643. How to think it and how to do it, particularly at lower cost. (Ken Kern has several other books from the same helpful perspective.)

Los Alamos National Laboratory. *Passive Solar Design Handbook, Vols. 2 and 3.* US Government Printing Office. These two books describe the Solar Load Ratio and Load Collector Ratio methods of estimating the winter performance of (high mass) passive solar buildings, and include the necessary tables for using those methods. Very tough going for the uninitiated. William Shurcliff (see below) has published a simplifying guide to Volume 3.

Mazria, Edward. *The Passive Solar Energy Book.* Rodale. Rules of thumb for standard passive systems. Includes sun charts and climate and weather tables.

Langdon, William K. *Movable Insulation.* Rodale. Manufactured and do-it-yourself night window insulation.

Shelton, Jay. *Solid Fuels Encyclopedia.* Garden Way. Everything you wanted to know about heating with wood.

Shurcliff, William A. *Air-to-Air Heat Exchangers for Houses.* William A Shurcliff, 19 Appleton Street, Cambridge, MA 02138. Guide to principles, performance, and cost, with information on many specific brands. Also includes discussion of pollutants, including a chapter on radon gas.

Shurcliff, William A. *Saunders' Shrewsbury House.* Entire book is about a 100 percent solar heated house in Massachusetts that incorporates many innovative features.

Shurcliff, William A. *Simplifying Guide to the "Vol. 3 Passive Solar Design Analysis."* It is still formidable.

Shurcliff, William A. *Superinsulated Houses and Double Envelope Houses.* Brick House. Early examples discussed.

Shurcliff, William A. *Thermal Shutters And Shades.* Brick House. Principles and specific details of strategies for reducing heat loss through windows.

Small Homes Council — Building Research Council, University of Illinois, One East Saint Mary's Road, Champaign, IL 61820. Many excellent short publications on house design and construction subjects.

Underground Space Center, University of Minnesota. *Earth Sheltered Residential Design Manual.* Van Nostrand Reinhold. Much information about underground and earth-sheltered housing, from experts in the field.

Walker, Les and Millstein, Jeff. *Designing Houses.* Overlook Press, Lewis Hollow Road, Woodstock, NY 12498. A step-by-step guide for prospective home owners who want to create a personal plan.

**Superinsulation:**

Argue, Robert and Marshall, Brian. *The Super-Insulated Retrofit Book.* Renewable Energy in Canada, 107 Amelia Street, Toronto, Canada M4X 1E5. Aware home-owners' guide to energy-efficient renovation.

McGrath, Ed. *The Superinsulated House.* That New Publishing Company, 1525 Eielson St., Fairbanks, AK 99701. Some practical advice and details from someone with experience in very cold climate housing.

**Greenhouses:**

Fisher, Rick and Yanda, Bill. *The Food And Heat Producing Greenhouse.* John Muir Publications, PO Box 613, Santa Fe, NM 87501. How to design and build a solar greenhouse, with numerous examples, from people with a lot of experience.

Klein, Miriam. *Horticultural Management of Solar Greenhouses in the Northeast.* The Memphramagog Group, PO Box 456, Newport, VT 05855. One of the best books on growing food in the solar greenhouse, with specific information on many plant varieties. Includes natural methods of pest control and fertilizing.

Nearing, Helen and Scott. *Building and Using Our Sun-Heated Greenhouse.* Garden Way. The experiences of two homesteaders in New England.

Ott, John N. *Health And Light.* Pocket Books. The effects of natural and artificial light on plants and people.

Smith, Shane. *The Bountiful Solar Greenhouse.* John Muir Publications, PO Box 613, Santa Fe, NM 87501. Recommended.

**Construction:**

Schwolsky, Rick and Williams, James I. *The Builder's Guide to Solar Construction.* McGraw-Hill. Includes theory and general design principles, as well as many construction details.

Syvanen, Bob. *Some Tricks of the Trade from an Old-Style Carpenter.* Bob Syvanen, 179 Underpass Road, Brewster, MA 02631

Syvanen, Bob. *Interior Finish.*

**Periodicals:**

*Convection Loops,* Box AF, Stanford, CA 94305. Monthly newsletter forum for designers of convective loop buildings.

*Energy Design Update,* PO Box 716, Back Bay Annex, Boston, MA 02117. Monthly newsjournal related to superinsulation, with practical construction tips, informative articles, and reports on new products, research findings, books, computer programs, etc.

These three below are useful not only for their articles but for their product advertising:

*Fine Homebuilding,* The Taunton Press, 52 Church Hill Road, Box 355, Newtown, CT 06470. With well-illustrated examples.

*New Shelter,* 33 E Minor Street, Emmaus, PA 18049.

*Solar Age,* Harrisville, NH 03450.

## Reference Section
## D
## SOURCES OF PRODUCTS

The "Whole Earth Catalog" is a good first place to look for information about a lot of useful things, including energy-related building products: PO Box 428, Sausalito, CA 94966. Two other catalog sources for solar energy products are: "Peoples' Solar Sourcebook," at $5 from Solar Usage Now, Bascom, OH 44809, and the $3 "Solar Catalog" of Solar Components Corporation, PO Box 237, Manchester, NH 03305.

Many energy-related products are advertised in *New Shelter* and *Solar Age* magazines.

### Finding the Right Product:

**Big Fin** absorber extrusions are attached to water tubing and mounted in a sunspace for preheating domestic hot water: Zomeworks, PO Box 712, Albuquerque, NM 87103.

**caulkings and sealants** are used at cracks, at vapor and infiltration barrier joints, and at glass edges. They should have at least a 20-year effectiveness; select material (probably acrylic or urethane) appropriate for the particular usage and read package labels. Acoustical sealants are non-hardening and are used only in non-exposed joints. Silicones do not seal well to wood. Foam may be most effective around window frames. Try building materials suppliers, or Tremco, 10701 Shaker Blvd, Cleveland, OH 44104.

**coating for glass and acrylic** to prevent formation of water droplets (as on greenhouse glazing): Solar Sunstill, Box W, Setauket, NY 11733.

**exterior doors:** foam-insulated steel doors are standard, with magnetic weatherstripping and weather-tight sills.

**fire alarm equipment:** If not available through local electricians write B-R-K Electronics, 780 McClure Street, Aurora, IL 60504.

**glaziers' backer rod** is used under sealants in uniform-width joints such as the space between a

metal window and a gypsum-board edge molding.
Available from glass installers.

**heat piston** for opening a greenhouse vent
automatically: Dalen Products, 11110 Gilbert Drive,
Knoxville, TN 37922; Heat Motors, PO Box 411,
Fair Oaks, CA 95628.

**hygrometer** for direct reading of relative humidity,
and maximum-minimum thermometers: Brooklyn
Thermometer Company, 90 Verdi Street,
Farmingdale, NY 11735.

**outrigger wall systems:** Information on the
Larsen Truss may be purchased from John Hughes,
–204 10830 107 Avenue, Edmonton, Alberta T5H
0X3; on the Leclair system from Gerald Leclair,
607 Fleet Avenue, Winnipeg R3M 1J7.

**stainless steel water tank** designed for domestic
hot water systems: Solar Survival, PO Box 275,
Cherry Hill Road, Harrisville, NH 03450.

**Tu-Tuff** is a very tear-resistant vapor barrier
material made from cross-laminated polyethylene
sheeting: Sto-Cote Products, Drawer 310,
Richmond, IL 60071.

**Tyvek** infiltration barrier (NOT a vapor barrier) is
effective at the outside of wall framing: available
from many building suppliers, or duPont Company,
Centre Road Building, Room 11K4, Wilmington, DE
19898.

**vapor barrier boxes** for electrical wiring: NRG
Saver Distributors, Box 50, Group 32, RR1B,
Winnipeg, Manitoba R3C 4A3.

**vapor barrier paint:** Insul-Aid, Glidden Company,
Yellow Pages.

**wall jacks** make it easier to raise assembled thick
wall systems: Proctor Products, 210 Eighth Street
South, Box F, Kirkland, WA 98033.

**windows:** compare infiltration ratings, U-values,
durability.

**windowsill glass stops:** aluminum extrusions
E-0927 or E-0929 by Tubelite, available from glass
suppliers or Tubelite, Box 118, Reed City, MI
49677.

**window insulation:** Thermocell Miniblinds are a
coated-mylar curtain of accordion-fold cells, folding
compactly, and effective to R-7. Thermal
Technology Corporation of Aspen, 601 Alter Street,
Broomfield, CO 80020.

212

## Reference Section
### E
### GREENHOUSE FOOD PRODUCTION IN COLD CLIMATES

(from conversations with New Hampshire growers Sara Cox-Kaufman, Bruce Kaufman, Sonia Wallman)

### OUTSTANDING PLANT VARIETIES

Below is a list of plants that are particularly recommended. (Have fun experimenting with your own choices too.) The plants listed meet some or all of the following criteria:

- grows well under greenhouse conditions
- highly productive for space used
- long season productivity
- hardy, withstands extremes of heat and cold
- versatile, many uses both raw and cooked
- high nutritional value

### Vegetables

**bunching onions**

**Swiss chard, rhubarb chard**

**celery:** always keep soil moist

**broccoli,** De Cicco or other variety with many sprouts

**Chinese cabbage,** any spring variety

**collards, kale:** outstandingly nutritious, outstandingly hardy

**leaf lettuce,** Black-seeded Simpson and Oakleaf

**Romaine lettuce,** Cosmos MC

**Kyona (Mizuna):** a very mild and productive mustard green

**Chinese Pac Choi:** sweet, resists bolting

**edible chrysanthemum** (Shungitu): highly productive, both leaves and flowers edible

**Curly Cress:** for flavoring; must be kept cut back, not allowed to go to seed

(All the above varieties are available from Johnny's Selected Seeds, Albion, ME 04910)

**tomatoes,** Sweet 100: a small cherry type, very sweet and abundant, grows as a vine
(Available from Harris Seeds, 3670 Buffalo Road, Rochester, NY 14624)

**cucumber,** Pandex (European type); self fruitful, does not need bees for pollination
(Available from Stokes Seeds, 737 Main Street, Box 548, Buffalo, NY 14240)

**radish,** Scarlet Globe forcing radish: bulbs develop even though days are short
(Source unknown; formerly available from Stokes)

**Herbs**

(Many other herbs can also be grown.)

**creeping rosemary:** does not need a rest period, can be transplanted from summer garden for continued use all winter

**mint:** grow in a hanging basket, as it spreads aggressively

**dwarf nasturtium:** a delight to see; both flowers and leaves are edible for salads

**parsley, flat leaf:** more productive and flavorful than curled parsley

## BRIEF SUGGESTIONS FOR GREENHOUSE MANAGEMENT

**Promote rapid growth;** don't let plant growth slow down. Use high quality compost for bedding soil, in beds 2-3 feet deep with good drainage at the bottom. Mix soil well and have it tested for possible deficiencies. Water thoroughly as needed, but not too frequently. (Drying out soil below 1/4" depth can kill tiny root hairs and slow plant growth.) Help keep soil warm by using warm water for watering, near 70-80F if possible. Feed plants twice a month with an organic fertilizer such as manure water, or with a foliar spray (available from Necessary Trading Company, Box 305, New Castle, VA 24127).

**Conserve space and save production time** by starting most plants in flats. Have new plants ready to place in beds when other plants are through growing and their space is available. Use vertical growing space with strings, stakes, and lattices for tomatoes, cucumbers, melons, etc., but be careful not to shade shorter plants.

**Have plants well established** before the cold, dark, short-day months of winter when plants grow slowly. Full-grown greens remain edible for a long time if night-time temperatures drop to 50-55F or lower (not below freezing). (High temperatures encourage greens to bolt — i.e. send up a flower stalk and lose edibility.)

**Avoid overheating as well as freezing.** Try to keep temperatures under 90F. Excessive heat can slow growth, prevent fruit formation, kill plants. Aphids multiply more rapidly at temperatures above 65-70F.

**Control humidity:** 50-60% RH is excellent; high humidity (over 75%) encourages mold which tends to lead to other problems. Be especially careful in cloudy humid weather. Water plants in early morning: midday watering could scorch plants, and evening watering leaves moisture on leaves overnight.

**Guard against insects and diseases:**

• **Adequate ventilation and light are of primary importance.** Good air circulation helps prevent mold and fungus diseases; bright light helps to discourage aphids (as well as promoting rapid and healthy growth).

• **Attack insect pests at first sight.** Be watchful: insects can multiply very rapidly. White flies and aphids are usually the biggest threats. Try biological controls. Ladybugs control aphids; spraying with Safer's Soap helps. Yellow-painted boards or hanging balls with Tanglefoot on them can trap white flies. Some can be put in corners to monitor the white fly population. If insects get out of hand, spray with Red Arrow, a rotenone-pyrethrum blend. Local agricultural agents may offer specific help. Products mentioned above (including biological controls such as ladybugs) are available from Necessary Trading Company (address above).

• **Keep greenhouse cleaned up:** remove fallen leaves to compost; destroy badly infected plants; don't leave debris that pests can hide in.

Gardening in a greenhouse is very different from gardening out-of-doors. Books on greenhouse gardening are listed under "Related Reading" in the reference section.

# Reference Section
## F
## HEAT AVAILABLE FROM FUELS

The cost of fuel per million Btu (MMBtu) of available heat is a convenient figure both for comparison of different fuels and for estimating yearly heating costs.

The chart below shows the amount of heat typically available from common fuels. The blank spaces on the right-hand side of the chart can be filled in with the current figures for your location to obtain the local cost per million Btu for each type of fuel. First, write in the applicable cost in the column "Local Unit Cost" (be sure you are using the right units), and then multiply by the factor shown; write the result in the column under "Local Cost per million Btu".

Example: if electricity costs $.05/kWhr, that can be written in under "Local Unit Cost" for electric heat, and then multiplied by 293 to find the cost of electric heat per million Btu (MMBtu), which would be $14.65. In this example, a yearly heating load of 8 MMBtu would cost about $117 for electric heat.

If the efficiency is known to be different from that shown above, adjust the figure you have obtained for cost per MMBtu by multiplying by the shown efficiency and then dividing by your efficiency.

The figure shown for typical fireplace efficiency (5%) is probably too high for many conventional fireplaces. An ordinary fireplace without a metal heat exchanger, with or without glass doors, often contributes a net heat loss even when burning hotly, because it takes so much heated room air up the chimney.

| Type of Heat | Typical Efficiency | Available Heat | Local Unit Cost | factor | Local Cost per million Btu |
|---|---|---|---|---|---|
| ELECTRIC resistance heat | 100% | 3,413 Btu/kWhr | $ _____ /kWhr | x 293 | = $ _____ /MMBtu |
| HARDWOOD airtight stove | 50% | 4,000 Btu/lb 12,000,000 Btu/cord | $ _____ /cord | x .0833 | = $ _____ /MMbtu |
| fireplace | 5% | 1,200,000 Btu/cord | $ _____ /cord | x .8333 | = $ _____ /MMBtu |
| OIL | 65% | 91,000 Btu/gal | $ _____ /gal | x 11 | = $ _____ /MMBtu |
| GAS natural gas | 65% | 65,000 Btu/therm | $ _____ /therm | x 15.38 | = $ _____ /MMBtu |
| liquid propane | 70% | 1,750 Btu/cf | $ _____ /cf | x 571 | = $ _____ /MMBtu |
| COAL | 60% | 8,000 Btu/lb | $ _____ /lb | x 125 | = $ _____ /MMBtu |

216

## Reference Section
## G
## SOLAR DESIGN CALCULATIONS

Parallel with the development of passive solar house design techniques in the past ten years has been the development of mathematical techniques for analyzing and predicting the performance of those designs. These range from simple hand calculations based on rules of thumb to very sophisticated programs running on large computers.

There are three major questions which solar design calculations can attempt to answer:

1. Will the design work? For example, a designer may want to know whether the thermal storage planned for a house is adequate, or if the house will overheat. If a proven design is used with little modification, this question would not normally arise.

2. Exactly how well will the house perform when it is occupied? Designers and owners may both want to know how much it will cost to heat the house each year, or just how hot or cold parts of it will get. This type of quantitative prediction is the most difficult question to answer with accuracy, since there are so many unknown factors which have a large effect on actual performance. Even the most advanced calculation methods implemented on large computers have a fairly poor record in predicting the actual performance of occupied houses, because such factors as infiltration, weather, and owner lifestyles are unpredictable.

3. What is the effect of making certain design changes? (or, How can a design be optimized?) This is the question which calculation methods are best equipped to answer, since it involves the relative effect of changes rather than absolute quantities. If a calculation method is not too difficult to use, several variations on a design can be tried in an attempt for optimization. This can be coupled with economic analysis to determine the cost-effectiveness of the modifications.

## Simplified Correlation Methods

Several methods have been developed for estimating the total yearly or monthly heating requirements of houses. One of the simplest is the degree-day method, as outlined in ASHRAE and other sources. This method can be very useful in giving a quick estimate for conventional houses, but cannot adequately handle most solar designs. To predict the performance of solar designs, various correlation methods have been developed for predicting the overall heating needs for the most common systems such as direct gain, thermal storage wall and attached greenhouse.

The most widely used passive solar design calculation methods are the Solar Load Ratio (SLR) and the Load Collector Ratio (LCR) methods developed by J Douglas Balcomb et al. at the Los Alamos Scientific Laboratory. These are correlation methods based on the results of many large computer simulations. They can be used by hand, and there are also programs available which implement them on programmable calculators and microcomputers, making them easier and quicker to use.

The SLR and LCR methods are described in Volumes 2 and 3 of the "Passive Solar Design Handbook", which also contains the tables necessary for using the methods. However, learning to use the methods from those books is a difficult and time-consuming process. The "Simplifying Guide" mentioned below can provide help, but probably the easiest way to learn these methods is with a good, well-documented microcomputer program.

Correlation methods such as SLR are limited in the type and complexity of design which they can handle, and they provide very little detail in their performance predictions. The relative complexity of many solar designs, including buffered designs, can mean that they are are not suitable for analysis with these correlation methods.

## Simulation

One solar calculation method which avoids many of the limitations of the correlation methods is simulation, or thermal network modeling. In this method, the temperatures of important locations in a design are predicted for frequent intervals (usually each hour) under fairly realistic conditions. Using this method the performance of a design can be observed under desired conditions (such as a harsh weather period) and the total heating needs can be estimated by running extended simulations using typical weather data for the location of the design. The flexibility of the simulation method can make it suitable for virtually any solar design system.

The enormous number of calculations which are required to simulate hourly temperatures and heat flows eliminates the possibility of using this method by hand, and some sort of computing device is necessary. However, the biggest drawback of the simulation method is the time required in learning to use it and the time that it can take to set up a mathematical model of a design; the accuracy of the results is completely dependent upon the accuracy of the mathematical model as defined by the person performing the analysis.

The simulation programs which allow the greatest detail and flexibility are the programs written for large computers. Because of cost or inconvenience these programs have not been widely used by solar house designers in the field, although they have been extensively used for research and for large design projects.

Simulation programs are also available for programmable calculators, and these can be attractive because of their low cost. Their big disadvantages are the limited detail and the fact that they are too slow to make extended or year-long simulations of a design practical. They are generally used only when fine-tuning a design, to observe the performance on particular design days.

The microcomputers which are finding their way into more and more offices and homes can provide a middle ground in sophistication, speed and cost between the large computer programs and the calculator programs. One big advantage that microcomputers have over calculators is the possibility of programs that lead the user (with instructions on the video screen) through the input process, making the program much easier to use. There are several simulation programs now available for microcomputers, and the falling cost of the necessary hardware makes this an option which is likely to become increasingly popular with solar designers.

Simulations for this book were made on Community Builders' SUNDESIGN microcomputer program by Jonathan and David Booth.

References:

DOE Passive Solar Design Handbook Volume 3 (Stock Number 061-600-00598-6)

(defines SLR and LCR methods, with all tables)

Superintendent of Documents
U S Government Printing Office
941 North Capitol Street NE
Washington, D C 20402
($12)

Simplifying Guide to the Los Alamos Vol. 3 Passive Solar Design Handbook

William A Shurcliff
19 Appleton St.
Cambridge, MA 02138
($9)

Microcomputer Methods for Solar Design and Analysis (SERI/SP-722-1127)

(Survey of available programs, may be updated again soon)

SERI Document Distribution Service
Solar Energy Research Institute
1617 Cole Boulevard
Golden, CO 80401
($4)

Solar Heating and Cooling Computer Programs (EPRI ER-1146)

(survey of programs, including manual and programmable calculator methods)

Research Reports Center
Electric Power Research Institute
Box 50490
Palo Alto, CA 94303

## Community Builders

Since 1954 Don Booth and others associated with Community Builders have enjoyed the challenge of creating attractive and comfortable homes free of maintenance problems. Since the sharp rise in fuel costs in the mid-1970s they have focused their attention on energy-efficient housing and have built a variety of types of solar homes, always seeking the most practical approaches.

In addition to building in the area around Canterbury, NH, Community Builders provides basic design and consultation services in a wider area. With their Sundesign computer analysis service they can do simulations of the thermal performance of houses that are still in the planning stage and help people evaluate various energy efficient design options.

Working drawings of some Community Builders' houses are available for purchase.

For information write Community Builders, Canterbury, NH 03224.

# INDEX